The Visual Guide to

EXTRA DIMEN -SIONS

Volume 2

The Physics of the Fourth and Higher Dimensions, Compactification, and Current and Upcoming Experiments to Detect Extra Dimensions

Chris McMullen, Ph.D.

The Visual Guide to Extra Dimensions

Volume 2: The Physics of the Fourth and Higher Dimensions, Compactification, and Current and Upcoming Experiments to Detect Extra Dimensions

Copyright (c) 2008 Chris McMullen

All rights reserved. This includes the right to reproduce any portion of this book in any form.

www.faculty.lsmsa.edu/CMcMullen

Custom Books

Nonfiction / science / physics / string theory

Nonfiction / science / mathematics / geometry

ISBN: 1441497536

EAN-13: 9781441497536

Volume 1
> Visualizing the Fourth Dimension, Higher-Dimensional Polytopes, and Curved Hypersurfaces

Volume 2
> The Physics of the Fourth and Higher Dimensions, Compactification, and Current and Upcoming Experiments to Detect Extra Dimensions

The Visual Guide to Extra Dimensions

Volume 2

The Physics of the Fourth and Higher Dimensions, Compactification, and Current and Upcoming Experiments to Detect Extra Dimensions

Chris McMullen, Ph.D.

Contents

Introduction   7

Dedications   9

Chapter 6: Higher-Dimensional Vectors
  6.0 Scalars and Vectors   11
  6.1 Vectors in the Second Dimension   14
  6.2 Vectors in the Third Dimension   18
  6.3 Vectors in Higher Dimensions   21
  6.4 Generalizing the Dot Product   23
  6.5 Generalizing the Cross Product   26
  6.6 Higher-Dimensional Matrices   32

Chapter 7: Force Fields
  7.0 Gravitational Forces   37
  7.1 Electromagnetic Forces   46
  7.2 Field Lines   53
  7.3 Gauss's Law   62
  7.4 Stokes's Theorem   68
  7.5 Higher-Dimensional Gravity   71
  7.6 Higher-Dimensional Electromagnetism   74
  7.7 Nuclear Forces   78

Chapter 8: Compactification
  8.0 Non-Euclidean Geometry   81
  8.1 Spacetime Curvature   85
  8.2 Hidden Dimensions   89
  8.3 Compactification Schemes   93
  8.4 Toroidal Compactification   100
  8.5 Kaluza-Klein Excitations   105
  8.6 Large Extra Dimensions   107
  8.7 Higher-Dimensional Models   113

Chapter 9: Experimental Searches
  9.0 Tests of Newton's Law of Gravitation   116
  9.1 Elementary Particles and Their Interactions   122
  9.2 High-Energy Collider Phenomenology   128
  9.3 Collider Searches for Extra Dimensions   143
  9.4 Astrophysical and Cosmological Constraints   159

Volume 2: The Physics of the Fourth and Higher Dimensions...

Chapter 10: Spacetime
   10.0 Special Relativity   162
   10.1 General Relativity   174
   10.2 Quantum Mechanics   186
   10.3 Quantum Field Theory   193
   10.4 Grand Unified Theories   195
   10.5 String Theory   202
   10.6 Supersymmetry and Superstrings   207
   10.7 M Theory   209

Volume 2 Contents   212

References and Further Reading   213

Puzzle Solutions   215

Author's Qualifications   218

Introduction

My fascination with extra dimensions began when I was in high school and encountered one of Rudy Rucker's books on the fourth dimension during one of my trips to the math and science shelves of a bookstore. I remember staying up into the wee hours contemplating the ana and kata directions, drawing hypercubes and hyperspheres, and solving the puzzles from Rudy Rucker's book.

I was fortunate to combine my interest in extra dimensions with my expertise in physics during my doctoral studies at Oklahoma State University: Some papers in the 1990's had stimulated interest in large string-inspired extra dimensions in the particle physics community. I earned my Ph.D. in physics when an explosion of professional research was published in this field, and have continued to publish papers on the collider phenomenology of large extra dimensions ever since.

Recalling my appreciation for Rudy Rucker's works, I wanted to make my own humble contribution to readers with an interest or background in math or physics who are curious about extra dimensions. I hope that my work will engage the interest of some readers who share my curiosity for the fourth and higher dimensions.

This first volume is dedicated toward a geometric extra dimension very much like the three known dimensions. It begins by surveying interesting features of 1D and 2D worlds, since we can understand many higher-dimensional concepts through analogies in the lower dimensions. Some common objections to studies of a fourth dimension are addressed, especially to convey the message that times have changed: There is good motivation for the possibility of superstring-inspired extra dimensions that are much larger than originally thought – large enough that current and upcoming experimental searches are underway with prospects for detecting them.

One of the main goals of the first volume is to develop techniques for visualizing extra dimensions; numerous illustrations, many of which are novel, aim to aid in this process. Another goal is to thoroughly analyze some fundamental 4D objects, especially the tesseract and the glome. The final chapter of this volume looks at a hypothetical hyperuniverse, in which these visualization techniques and basic geometric objects are applied. This chapter serves as a useful bridge between the higher-dimensional geometry considered in the first volume and the higher-dimensional physics developed in the second volume.

Volume 2: The Physics of the Fourth and Higher Dimensions…

This book is primarily conceptual, for the benefit of readers who do not have a strong background in the mathematics of particle physics or superstring theory, yet there is a mathematical component to this book, since it is anticipated that most readers will have some interest or background in mathematics. It will be desirable to have a good handle on the fundamentals of algebra, geometry, and graphing. Where it is deemed useful to discuss more advanced mathematics, higher-level concepts are developed from these starting levels.

Nonetheless, more emphasis is placed on the concepts than on the math, since this is intended to be an enjoyable book on an interesting topic, which should not read like a textbook. This book is very detailed and technical conceptually in an effort to help stimulate and engage the interest of such mathematically-minded readers.

Several illustrations are intended to challenge readers visually. For example, several illustrations combine together to show a hypercube unfold, the various 3D cross sections of a hypercube are drawn, the 3D projection is depicted for a rotating hypercube, higher-dimensional cylinders and tori are graphed, and a higher-dimensional staircase is drawn. Puzzles scattered throughout the book offer additional challenges.

The second volume looks at higher-dimensional mathematics, higher-dimensional force laws related to Gauss's law, the important issue of compactification, current and upcoming experimental searches for extra dimensions, and a little background in spacetime, quantum mechanics, and string theory.

May you enjoy the book as much as I enjoyed reading my first book on extra dimensions.

Dedications

I would like to thank the many teachers who inspired my interest in math and science, my passion for teaching, and many of the good qualities that I strive for in my teaching, research, mentorship, and life. I particularly appreciate those teachers who were excellent motivators and mentors and those whose courses were very challenging, rigorous, and thought-provoking. One of my favorites is Dr. Robert Chianese, who taught a fascinating course on Cold War Literature – so much so that I have since re-read many of the assigned readings, putting forth much more effort after the credit had already been earned. I had the good fortune of taking a few courses with Dr. Duane Doty, who is a superb mentor, motivator, and teacher. My Master's thesis, under the guidance of Dr. Miroslav Peric, immensely helped to develop my confidence in challenging myself with research. Teachers such as Dr. A.C. Cadavid, Dr. Peter Collas, and Dr. Robert Park helped to instill my interest in rigorous mathematics, and inspired the high level of mathematics that I strive to incorporate into the courses that I teach. Dr. K.S. Babu is the ideal model for a command of knowledge in one's area of expertise. I am very grateful for the opportunities that I have had to collaborate with my mentor, Dr. S. Nandi, on multiple research projects, in addition to being a student in the excellent courses that he taught. I would also like to mention my high school geometry teacher, Mr. Ratkovic, who is not only an exceptional teacher, but who identified gifted students in his classes and found ways to engage their interest and challenge their minds while still teaching to the rest of the class. I appreciate that so much that it is one of my primary reasons that I accepted my current position at the Louisiana School for Math, Science, and the Arts, and it serves as a strong source of my motivation to challenge young minds in math and physics.

I must also thank my family for their encouragement and always believing in my ability, including my mom, dad, grandma, and aunt. I am also very grateful for the invaluable support of my wife, whose Master's thesis was also on the subject of collider phenomenology of large extra dimensions.

I dedicate this book to all those who have had a positive impact on my life.

Volume 2: The Physics of the Fourth and Higher Dimensions…

6 Higher-Dimensional Vectors

We will begin our discussions of the physics of one or more extra dimensions by first considering how a fourth Euclidean extra dimension would impact standard physical formulas and laws; we will save compactification until Chapter 8. Most physical quantities behave as scalars or vectors, depending upon whether or not they have direction, while other physical quantities can be represented as matrices. It turns out that some physical quantities that we perceive to be vectors in 3D space would actually be represented with matrices in higher (or even lower) dimensions. This point will be particularly important when we consider the fundamental laws of physics in the next chapter. A reader who is well-versed in physics or mathematics may find some of the material in this chapter to be trivial, but this chapter serves as a necessary prelude in order to make some of the concepts of Chapter 7 accessible to readers who are not familiar with second-rank tensors (which we will simply refer to as matrices in this more accessible account, although there is an important distinction).

6.0 Scalars and Vectors

A physical quantity is something that can be measured, such as length or time. There are seven fundamental measurements: length, time, mass, temperature, electric current, luminous intensity, and amount of substance (or mole number). Any other physical quantity can be determined from these basic measurements. For example, speed can be computed from measurements of length and time.

Some physical quantities have direction, while others do not. When measuring force, for example, it is possible to note not only how much force there is, but also in what direction the force is pulling. Thus, force is a physical quantity that has direction. It does not make sense, on the other hand, to ascribe a sense of direction to a temperature measurement. For the most part, a physical quantity that has an inherent sense of direction is termed a *vector*, while a physical quantity for which direction does not make sense is called a *scalar*.

> **In a basic sense, a vector is a physical quantity that has both a magnitude and direction, while a scalar has just a magnitude.**

Measurement of a scalar requires measuring a single value, called the *magnitude*. The magnitude of a quantity is a numerical value with units (such at 10 centimeters or 25 kg) that indicates how much of a given quantity there is. Mass, energy, and temperature are examples of common scalars.

The magnitude of a quantity is a numerical value that indicates how much of that quantity there is.

Measurement of a vector entails measuring two values – magnitude and direction (e.g. 600 Newtons downward). Measuring a scalar is like asking *How much?* while measuring a vector is like asking *How much and which way?* Velocity, force, and momentum are examples of common vectors.

Speed and velocity are two similar terms. The distinction is that velocity is a vector, and thus includes direction, whereas speed does not. For a car that has a velocity of 70 mph east, its speed is simply 70 mph.

Most measures of length, such as the length of a metal rod, are scalars, but there are situations in which there may be a direction to associate with a distance. For example, when measuring how far a jogger runs, it can also be worth noting which way the jogger has moved. The total distance traveled is a scalar – simply how far the jogger has run all together – while the net displacement is a vector – i.e. it is a directed distance. If a jogger starts out heading north along a circular track, then after completing half a lap the total distance traveled is half the circumference, while the net displacement is one diameter to the west.

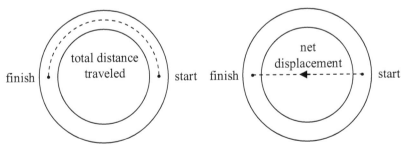

a jogger runs halfway around a circular track

Mass and weight are two similar terms, often confused since one is a multiple of the other (a factor of gravitational acceleration). One fundamental distinction is that one is a scalar and the other is a vector. The mass of an object is a measure of how easy it is to accelerate the object (i.e. to change its velocity), whereas the weight of an object is the gravitational force that a planet exerts on it. Mass is a scalar since it is equally difficult to accelerate an object in any direction (i.e. the amount of force needed, all together, does not depend on which way it is pulled). For example, on level ground with no wind, it is equally difficult to transport a 50-lb. rock 20 m to the east as it is 20 m to the north (neglecting insignificant factors such as the earth's rotation). Weight, on the other hand, has a definite direction – toward the center of gravity of a planet. The SI unit of mass is the kilogram (kg), while the SI unit of weight is the Newton (N).

The arithmetic of scalars and vectors is fundamentally different. Scalars add and subtract, for example, according to the rules of basic arithmetic, but vectors do not. Here are examples of scalar addition: If a car drives for 20 miles and then for 30 miles, the total distance traveled is 50 miles; twice as much force is required to accelerate a 100-kg box as a 50-kg box. Following are examples of vector addition: When a car drives 100 miles east and then 50 miles west, although the total distance traveled is 150 miles, the net displacement is 50 miles to the east; when a box is pushed east with a force of 30 N and north with a force of 40 N, the net force is 50 N (found by the Pythagorean theorem) and the direction is a somewhat more north than northeast. When adding or subtracting vectors, it is very important to factor into the effects of direction.

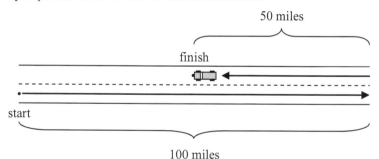

a car drives 100 miles east, then 50 miles west

Volume 2: The Physics of the Fourth and Higher Dimensions...

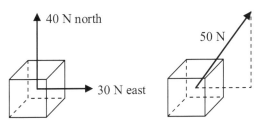

a box is pulled 30 N east and 40 N north

6.1 Vectors in the Second Dimension

It is useful to represent a vector by an arrow (↑), where the length of the arrow corresponds to the magnitude.

> **Notation**: Vectors and scalars can be distinguished by placing an arrow above a quantity that is a vector (e.g. velocity $\vec{v}$), but not for a scalar (e.g. speed v). That is, the symbol $\vec{v}$ means how fast and which way, whereas v only means how fast. It is also conventional to italicize symbols that represent scalar quantities, such as time t, speed v, and work W, but not to italicize units such as meters m, seconds s, and kilograms kg.

There is conceptual power in realizing that a vector can be moved around. This is not true of graphs in general. For example, if a line is moved, the equation for the line changes. However, a vector can be moved anywhere and still be the same vector provided that the length and direction of the vector are unchanged. It is often much easier to visualize vector addition, for example, by moving one or both vectors around. The tail of a vector is not fixed to one location.

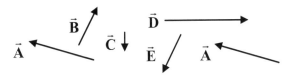

some vectors; the first and last vectors are the same

A 2D vector can be resolved into two independent coordinates – e.g. horizontal and vertical components. With respect to a Cartesian

14

coordinate system, it has x- and y-components (not to be confused with x- and y-coordinates). The x- and y-components can be found by projecting the vector onto the x- and y-axes, respectively. The projection of the vector onto the x-axis can be visualized by shining parallel rays of light in the negative y-direction; the shadow cast on the x-axis then represents the x-component of the vector. A similar shadow can be cast for the y-component. Alternatively, a right triangle can be formed with the vector as the hypotenuse and the components as the sides, where the sides are parallel to the x- and y-axes. The x- and y-components of a vector are denoted by using x and y subscripts (e.g. vector $\vec{A}$ has components A_x and A_y). Either or both components may be negative, depending on the direction of the vector.

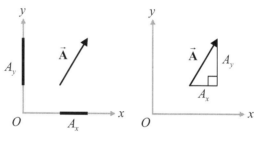

components of a vector

The x- and y-components of a vector indicate how far the vector extends along the x- and y-axes.

Two vectors $\vec{A}$ and $\vec{B}$ added together form a resultant vector $\vec{R}$. The resultant vector $\vec{R}$ acting alone is equivalent to the two vectors $\vec{A}$ and $\vec{B}$ acting together. For example, suppose that $\vec{A}$ is a displacement vector from Los Angeles to Dallas and $\vec{B}$ is a displacement vector from Dallas to New York City. The pilot of an airplane could fly along vector $\vec{A}$ and then vector $\vec{B}$, departing from Los Angeles and arriving in New York City via a layover in Dallas. Alternatively, the pilot could fly along the resultant vector $\vec{R}$, traveling directly from Los Angeles to New York City in a straight line. The resultant vector $\vec{R}$ is equivalent to $\vec{A}$ and $\vec{B}$ added together in the sense that either way the end result is the same – i.e. the initial and final points are the same (only the path is different). Symbolically, this equivalence is expressed as $\vec{A}+\vec{B}=\vec{R}$. This is called vector addition. It differs from the ordinary addition of numbers in that direction must be factored in. The magnitudes of the displacement vectors do not add this way – i.e. $A+B \neq R$ (unless $\vec{A}$ and $\vec{B}$ are parallel). The magnitudes of these displacement vectors are the distances traveled in each flight: A is 1251 miles, B is 1373 miles, and R is 2462 miles.

Two vectors $\vec{A}$ and $\vec{B}$ can be added graphically to find the resultant vector $\vec{R}$ by joining vectors $\vec{A}$ and $\vec{B}$ tip-to-tail – i.e. the tip of one connects to the tail of the other; the resultant vector $\vec{R}$ is then drawn from the tail of the first to the tip of the last. It does not matter whether the tail of $\vec{A}$ is joined to the tip of $\vec{B}$ or vice-versa since the two possibilities form a parallelogram; the resultant vector $\vec{R}$ is the diagonal of the parallelogram.

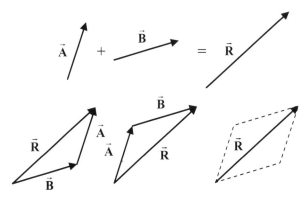

vectors add tip-to-tail, forming a parallelogram

The resultant of two vectors is equivalent to each of the vectors acting in conjunction in the sense that the resultant starts and finishes in the same place as the pair of vectors.

When two 2D vectors $\vec{A}$ and $\vec{B}$ are added to form a resultant vector $\vec{R}$, the components of $\vec{A}$ and $\vec{B}$ add up to the respective components of $\vec{R}$ – i.e. $A_x + B_x = R_x$ and $A_y + B_y = R_y$. Thus, vector addition $\vec{A}+\vec{B}=\vec{R}$ means to add the components according to $A_x+B_x=R_x$ and $A_y+B_y=R_y$, but does not mean to add the magnitudes (since $A+B \neq R$ unless $\vec{A}$ and $\vec{B}$ are parallel).

If two vectors $\vec{A}$ and $\vec{B}$ are perpendicular to each other, the magnitude R of the resultant vector $\vec{R}$ can be computed from the Pythagorean theorem since, in this case, $\vec{A}$, $\vec{B}$, and $\vec{R}$ form a right triangle: $R^2=A^2+B^2$. For example, if $\vec{A}$ has a magnitude of 3 N and points along the positive x-axis and $\vec{B}$ has a magnitude of 4 N and points along the positive y-axis, then the resultant vector $\vec{R}$ has a magnitude of $\sqrt{(3\,\text{N})^2+(4\,\text{N})^2}=5\,\text{N}$. (In the more general case where $\vec{A}$ and $\vec{B}$ are not perpendicular, the magnitude of $\vec{R}$ can be found by applying trigonometry.)

$$\vec{A} \longrightarrow + \vec{B} \uparrow \;=\; \vec{R} \nearrow \quad \vec{B}\uparrow \atop \vec{A}\to \qquad C\diagup \atop A$$

$$\vec{R}=\vec{A}+\vec{B} \qquad\qquad C^2=A^2+B^2$$

perpendicular vectors and the Pythagorean theorem

The magnitude A of a vector $\vec{A}$ is related to its components A_x and A_y through the Pythagorean theorem; the vector $\vec{A}$ and its components A_x and A_y always form a right triangle since the x- and y-axes are perpendicular.

Puzzle 6.1: How is possible for two *different* vectors to have a magnitude of 5 m and an x-component of 4 m?

Puzzle 6.2: How can three vectors with equal magnitude, but different directions, have a resultant of zero?

17

Volume 2: The Physics of the Fourth and Higher Dimensions...

Puzzle 6.3: Four vectors extend from the center of a regular pentagon to one of the five corners of the pentagon (so one corner is unused). If each vector has a magnitude of 1 m, what are the magnitude and direction of the resultant of these four vectors?

A unit vector is a vector with a magnitude of one unit. Two particularly useful unit vectors are the Cartesian unit vectors $\hat{i}$ and $\hat{j}$: $\hat{i}$ is a unit vector that points along the positive x-axis, and $\hat{j}$ is a unit vector pointing along the positive y-axis. The caret (^) above the unit vector that distinguishes it from a more general vector is often called a *hat*. Thus, $\hat{i}$ is often called "i-hat" and $\hat{j}$ called "j-hat." Any 2D vector $\vec{A}$ can be expressed in terms of Cartesian unit vectors in the form $\vec{A} = A_x \hat{i} + A_y \hat{j}$. Taking $\hat{i}$ to be east and $\hat{j}$ to be north, directions to a gas station could be given as $4\hat{i} - 4\hat{j}$ blocks, meaning go 4 blocks east and 4 blocks south. Alternatively, the directions could be to head southeast, but go $4\sqrt{2}$ blocks. Here, +4 blocks is the x-component of $\vec{A}$ and −4 is the y-component of $\vec{A}$; stating the components is equivalent to stating the magnitude and direction, which are $4\sqrt{2}$ blocks and southeast. Either path leads to the gas station.

A unit vector is a vector with a magnitude of one unit.

Puzzle 6.4: What is the magnitude of the vector $\vec{A} = -5\hat{i} + 12\hat{j}$?

Puzzle 6.5: What is the magnitude of the resultant of the vectors $\vec{A} = 2\hat{i} + 5\hat{j}$ and $\vec{B} = 4\hat{i} - 13\hat{j}$?

6.2 Vectors in the Third Dimension

In 2D, a vector can be specified by stating two independent quantities – either the x- and y-components or the magnitude and direction. In 3D, a vector consists of three independent components, the third being the z-component. A 3D vector is fully specified by its three components. Alternatively, it can be specified in terms of its magnitude and two angles to indicate its direction.

The direction of a vector lying in the xy plane can be given by a single angle – say, counterclockwise from the positive x-axis. In this convention, a vector with a direction of 0° points along the positive x-axis, a vector with a direction of 45° points along the line of symmetry $x = y$, a vector with a direction of 90° points along the positive y-axis, etc. In 3D, it is necessary to give two angles to specify the direction of a vector. A popular convention is to state the polar angle θ, which indicates how much the vector is tilted from the positive z-axis, and the azimuthal angle φ, which is measured counterclockwise (when viewed from the positive z-direction) from the positive x-axis after first projecting the vector onto the xy plane.[1]

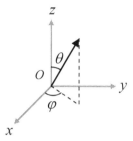

the direction of a 3D vector is specified by two angles

Consider a 2D vector lying in the xy plane that rotates about the z-axis. As it rotates, the vector sweeps out a circle. Now consider a 3D vector. For a fixed value of θ, varying φ rotates the vector through a cone; its tip rotates in a circle of latitude. For a fixed value of φ, on the other hand, varying θ rotates the vector through a great circle; its tip rotates in a circle of longitude. Thus, it is seen that the polar angle θ corresponds to latitude for a sphere, while the azimuthal angle φ corresponds to longitude. These angles θ and φ, along with the distance from the origin r, form a set of independent coordinates called *spherical coordinates* – an alternative to Cartesian coordinates especially useful for problems with spherical symmetry.

---

[1] While this notation is popular in physics, it is common in mathematics to use the symbol theta (θ) for the azimuthal angle and phi (φ) for the polar (or zenith) angle.

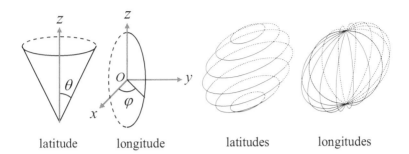

| latitude | longitude | latitudes | longitudes |

Puzzle 6.6: Describe a vector for which $\theta = 180°$. Repeat for $\theta = 90°$, and again for $\varphi = 90°$.

A 3D vector has three components A_x, A_y, and A_z. The unit vector along the positive z-axis is $\hat{\mathbf{k}}$. A 3D vector can be expressed in terms of Cartesian unit vectors as $\vec{\mathbf{A}} = A_x\hat{\mathbf{i}} + A_y\hat{\mathbf{j}} + A_z\hat{\mathbf{k}}$.

In 2D, the Pythagorean theorem relates the sides of a right triangle to its hypotenuse – i.e. $C^2 = A^2 + B^2$, where C is the hypotenuse and A and B are the legs. Put another way, the Pythagorean theorem is useful for finding the length of the diagonal of a rectangle. Generalizing the Pythagorean theorem to 3D, $D^2 = A^2 + B^2 + C^2$, where D is the length of the body diagonal of a cuboid and A, B, and C are the edge lengths of the cuboid.

The formula for the length of the body diagonal of a cuboid, which is a generalization of the Pythagorean theorem to 3D, can be derived as follows. The length of the face diagonal of the cuboid with edge lengths A and B, according to the usual 2D Pythagorean theorem, is $\sqrt{A^2 + B^2}$. The face diagonal with length $\sqrt{A^2 + B^2}$ is perpendicular to the remaining edge with length C. Thus, the face diagonal with length $\sqrt{A^2 + B^2}$ can be combined with the edge length C using the usual 2D Pythagorean theorem. The result is the 3D Pythagorean theorem, $D^2 = A^2 + B^2 + C^2$, for the length of the body diagonal.

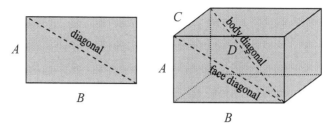

generalizing the Pythagorean theorem to 3D

The components A_x, A_y, and A_z of a 3D vector form the edges of a cuboid. Thus, the magnitude of a 3D vector is given by the equation $A = \sqrt{A_x^2 + A_y^2 + A_z^2}$.

Puzzle 6.7: What is the length of the body diagonal of a 9 m × 9 m × 9 m cube?

6.3 Vectors in Higher Dimensions

An N-dimensional vector has N Cartesian components, A_x, A_y, A_z, A_w, etc. If N is sufficiently large, it is more convenient to call the components $A_1, A_2, \ldots, A_N$ and the coordinates $x_1, x_2, \ldots, x_N$. For example, x_3 is z and A_2 is A_y.

The direction of an N-dimensional vector can be specified in terms of $N-1$ angles, including the azimuthal angle φ and polar/hyperpolar angles $\theta_1, \theta_2, \ldots, \theta_{N-2}$. For example, three angles are needed to specify the direction of a 4D vector: θ_2 is the angle between the vector and the w-axis, θ_1 is the angle between the z-axis and the projection of the vector onto the xyz hyperplane, and φ is the angle counterclockwise from the positive x-axis (as viewed from the positive zw plane) to the projection of the vector onto the xy plane.

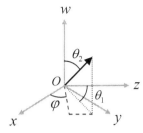

the direction of a 4D vector is specified by three angles

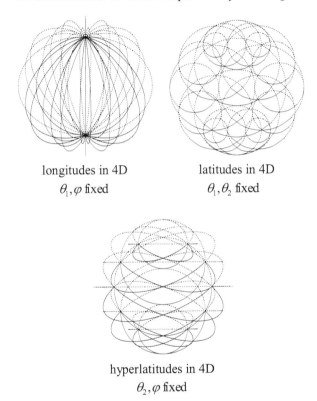

longitudes in 4D
θ_1, φ fixed

latitudes in 4D
θ_1, θ_2 fixed

hyperlatitudes in 4D
θ_2, φ fixed

Varying one angle while holding the other $N-2$ angles constant, an N-dimensional vector sweeps out a circle of longitude, latitude, or one of various types of hyperlatitudes. For example, in 4D, a great circle of longitude is swept out by fixing φ and θ_1, but varying θ_2; a circle of latitude is swept out by fixing θ_1 and θ_2, but varying

φ; and a circle of hyperlatitude is swept out by fixing θ_2 and φ, but varying θ_1.

The long diagonal of an N-dimensional orthotope with edge lengths $A_1, A_2, \ldots, A_N$ has length given by the formula $L = \sqrt{A_1^2 + A_2^2 + \cdots + A_N^2}$. For example, the long diagonal of a 4D orthotope with edge lengths A, B, C, and D is $L = \sqrt{A^2 + B^2 + C^2 + D^2}$.

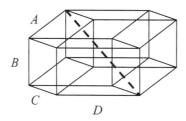

the Pythagorean theorem generalized to 4D

Puzzle 6.8: What is the length of the long diagonal of a 2 m × 2 m × 2 m × 2 m tesseract? What is the hypervolume of this tesseract? What is the 3D volume of its hypersurface?

There are three common, useful ways to multiply 3D vectors: the dot product, the cross product, and the outer product. Many fundamental equations that describe the physics of the known universe involve the dot product and cross product (and, less commonly, the outer product).

6.4 Generalizing the Dot Product

The dot product (aka scalar product or inner product) between two vectors results in a scalar – i.e. the dot product is just a number with no direction, although the vectors being multiplied do have direction. The dot product between two vectors $\vec{A}$ and $\vec{B}$ is written as $\vec{A} \cdot \vec{B}$ and equals the sums of the products of the respective components – e.g. in 2D, $\vec{A} \cdot \vec{B} = A_x B_x + A_y B_y$, and in 3D, $\vec{A} \cdot \vec{B} = A_x B_x + A_y B_y + A_z B_z$. The dot product can also be expressed trigonometrically as $\vec{A} \cdot \vec{B} = AB\cos\theta$, where A and B are the magnitudes of $\vec{A}$ and $\vec{B}$ and θ is the angle between them.

The basic trig functions – sine, cosine, and tangent – correspond to the ratio of sides of a right triangle. Relative to an angle θ in a right triangle with sides a, o, and h – where h is the hypotenuse, a is the side adjacent to θ, and o is the side opposite to θ – the basic trig functions are defined as follows: The sine of θ, expressed $\sin\theta$, which reads "the sine of theta," equals o/h; the cosine of θ, expressed $\cos\theta$, which reads "the cosine of theta," equals a/h; and the tangent of θ, expressed $\tan\theta$, which reads "the tangent of theta," equals o/a.

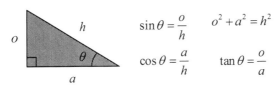

the basic trig functions

The dot product is used in physics, for example, in: the definition of work, which serves to define energy; Gauss's law, which is very fundamental to Newton's law of universal gravitation and Coulomb's law for the attraction or repulsion of electric charges; and computing the divergence of a field (e.g. gravity, electric, or magnetic), which indicates how much the field lines diverge at a given point.

In physics, work is defined in terms of force and displacement. Work is done if a force acts on an object and contributes toward the object's displacement. For constant force $\vec{F}$ and linear displacement $\vec{d}$, the work done W is the dot product between $\vec{F}$ and $\vec{d}$: $W = \vec{F} \cdot \vec{d} = Fd\cos\theta$. Here, θ is the angle between the force and the displacement. Work is the distance times the component of the force along the displacement ($F\cos\theta$ being the component of $\vec{F}$ along $\vec{d}$).

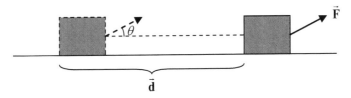

work is done by a force pulling a box

The dot product between vectors $\vec{A}$ and $\vec{B}$ depends on the angle θ between $\vec{A}$ and $\vec{B}$: $\vec{A} \cdot \vec{B} = AB\cos\theta$. Noting that $\cos 0° = 1$, $\cos 90° = 0$, and that $\cos\theta$ diminishes from 1 to 0 as θ increases from 0° to 90°, the dot product is maximal when $\vec{A}$ and $\vec{B}$ are parallel and zero when they are perpendicular. In the case of 0°, all of $\vec{A}$ projects onto $\vec{B}$, while in the case of 90°, $\vec{A}$ does not have a projection onto $\vec{B}$.

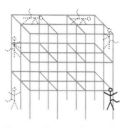

Puzzle 6.9: Near the surface of earth, where gravitational acceleration is 9.81 m/s², a 20-kg monkey climbs 8 m up the side of a jungle gym, crawls 8 m horizontally along the top, and then climbs 8 m down the other side. How much work did the monkey do against gravity?

Since $\cos 0° = 1$, the dot product of any vector with itself results in the square of its magnitude: $\vec{A} \cdot \vec{A} = A^2$. Since $\cos 90° = 0$, two vectors $\vec{A}$ and $\vec{B}$ are orthogonal (i.e. $\theta = 90°$, or perpendicular) if their dot product is zero.

The dot product is commutative, $\vec{A} \cdot \vec{B} = \vec{B} \cdot \vec{A}$, and distributive, $\vec{A} \cdot (\vec{B} + \vec{C}) = \vec{A} \cdot \vec{B} + \vec{A} \cdot \vec{C}$. Associativity, however, cannot be expressed in quite the same way as ordinary multiplication, $a(bc) = (ab)c$, since something like $\vec{A} \cdot \vec{B} \cdot \vec{C}$ is sheer nonsense (why?). However, associativity can be expressed using one scalar and two vectors: $a(\vec{B} \cdot \vec{C}) = (a\vec{B}) \cdot \vec{C}$. The vector algebra is particularly important in the context of extra dimensions, especially since the cross product is well-known to only be physically meaningful in 3D (but, fortunately, there is an alternative).

The dot product naturally generalizes to higher-dimensional spaces, as does its vector algebra. In 4D, for example, $\vec{A} \cdot \vec{B} = A_x B_x + A_y B_y + A_z B_z + A_w B_w$, and it remains true that

$\vec{A} \cdot \vec{B} = AB\cos\theta$. The fact that the dot product is both valid and physically meaningful in higher dimensions means that any laws of physics in the known universe that involve the dot product may also hold in a similar form in a higher-dimensional universe.

6.5 Generalizing the Cross Product

The cross product (aka vector product) between two vectors results in a vector whereas the dot product results in a scalar. Thus, the cross product is a fundamentally different kind of vector multiplication from the dot product. The cross product between two vectors $\vec{A}$ and $\vec{B}$ is written with a cross $(\vec{A} \times \vec{B})$ whereas the dot product is written with a dot $(\vec{A} \cdot \vec{B})$. The cross product in 3D is traditionally constructed in a seemingly unnatural way in terms of a determinant where the first row of the determinant contains the unit vectors $\hat{i}$, $\hat{j}$, and $\hat{k}$, while the second and third rows consist of the components of the first vector $\vec{A}$ and second vector $\vec{B}$, respectively:

$$\vec{A} \times \vec{B} = \begin{vmatrix} \hat{i} & \hat{j} & \hat{k} \\ A_x & A_y & A_z \\ B_x & B_y & B_z \end{vmatrix}$$

An $M \times N$ array of numbers – i.e. M rows and N columns – is called a *matrix*. Notationally, a matrix is denoted by enclosing the array in large parentheses. Examples of matrices:

$$\begin{pmatrix} 1 & 4 \\ -2 & 3 \end{pmatrix} \qquad \begin{pmatrix} 1 & 2 & 1 \\ 2 & 1 & 2 \\ 0 & 3 & 0 \end{pmatrix} \qquad \begin{pmatrix} 1 & 0 & 0 \\ 0 & 2 & 3 \end{pmatrix}$$

A matrix is an array of numbers.

For a matrix $\mathbf{A}$ (boldface for a matrix, but, unlike a vector, no arrow), its determinant is denoted by $\det(\mathbf{A})$ or $|\mathbf{A}|$. The determinant of a 2×2 matrix is found by multiplying the (1,1) and (2,2) elements together (going diagonally down and right), then multiplying the (2,1) and (1,2) elements together (going diagonally up and right), and then subtracting the latter product from the former product; where the notation (i, j) is used to denote the ith row and j^{th} column:

$$\det\begin{pmatrix} A_{11} & A_{12} \\ A_{21} & A_{22} \end{pmatrix} = \begin{vmatrix} A_{11} & A_{12} \\ A_{21} & A_{22} \end{vmatrix} = A_{11}A_{22} - A_{21}A_{12}$$

Example 6.0: Find the determinant of the following 2×2 matrix:

$$A = \begin{pmatrix} 1 & 3 \\ 2 & 8 \end{pmatrix}$$

Solution: First, going diagonally down to the right, $1 \cdot 8 = 8$. Next, going diagonally up to the right, $2 \cdot 3 = 6$. Putting these together, $\det \mathbf{A} = 1 \cdot 8 - 2 \cdot 3 = 2$.

The determinant of an $N \times N$ matrix can be expressed in terms of $(N-1) \times (N-1)$ matrices via the method of cofactors. The elements of the first row of a matrix can serve as the cofactors. For each cofactor, block out its column and row to obtain an $(N-1) \times (N-1)$ submatrix. The determinant of the $N \times N$ matrix equals the first cofactor (starting from the leftmost column, then moving right) times the determinant of its submatrix minus the second cofactor times the determinant of its submatrix plus the third cofactor times the determinant of its submatrix, and so on, continuing to alternate signs.

Example 6.1: Find the determinant of the following 3×3 matrix:

$$A = \begin{pmatrix} 1 & 2 & -1 \\ 6 & 1 & 0 \\ 3 & 4 & 5 \end{pmatrix}$$

Solution: The cofactors are the elements of the first row: $A_{11} = 1$, $A_{12} = 2$, and $A_{13} = -1$. Each cofactor creates a corresponding submatrix:

$$\begin{pmatrix} 1 & \blacksquare & \blacksquare \\ \blacksquare & 1 & 0 \\ \blacksquare & 4 & 5 \end{pmatrix} \quad \begin{pmatrix} \blacksquare & 2 & \blacksquare \\ 6 & \blacksquare & 0 \\ 3 & \blacksquare & 5 \end{pmatrix} \quad \begin{pmatrix} \blacksquare & \blacksquare & -1 \\ 6 & 1 & \blacksquare \\ 3 & 4 & \blacksquare \end{pmatrix}$$

The determinant of the 3×3 matrix is:

$$\det(\mathbf{A}) = \begin{vmatrix} 1 & 2 & -1 \\ 6 & 1 & 0 \\ 3 & 4 & 5 \end{vmatrix} = 1\begin{vmatrix} 1 & 0 \\ 4 & 5 \end{vmatrix} - 2\begin{vmatrix} 6 & 0 \\ 3 & 5 \end{vmatrix} + (-1)\begin{vmatrix} 6 & 1 \\ 3 & 4 \end{vmatrix}$$

$$\det(\mathbf{A}) = 1(1 \cdot 5 - 4 \cdot 0) - 2(6 \cdot 5 - 3 \cdot 0) + (-1)(6 \cdot 4 - 3 \cdot 1)$$

$$\det(\mathbf{A}) = 5 - 60 - 21 = -76$$

For the 3×3 determinant, there is an alternative to the method of cofactors. This alternative involves copying the first two columns and adding them at the end of the matrix, and multiplying down and up three diagonals, similar to the 2×2 determinant:

$$\begin{vmatrix} 1 & 2 & -1 \\ 6 & 1 & 0 \\ 3 & 4 & 5 \end{vmatrix} \begin{matrix} 1 & 2 \\ 6 & 1 \\ 3 & 4 \end{matrix}$$

$$= 1 \cdot 1 \cdot 5 + 2 \cdot 0 \cdot 3 + (-1) 6 \cdot 4 - 3 \cdot 1(-1) - 4 \cdot 0 \cdot 1 - 5 \cdot 6 \cdot 2$$
$$= 5 - 24 + 3 - 60 = -76$$

However, the advantage of the cofactor method is that it generalizes naturally to higher dimensions, whereas the diagonal method does not.

Example 6.2: Find the cross product of $\vec{\mathbf{A}} = 2\hat{\mathbf{i}} - \hat{\mathbf{j}} + \hat{\mathbf{k}}$ and $\vec{\mathbf{B}} = 3\hat{\mathbf{i}} + 4\hat{\mathbf{j}} - 3\hat{\mathbf{k}}$.

Solution: To find $\vec{\mathbf{A}} \times \vec{\mathbf{B}}$, form a 3×3 determinant with the unit vectors in the first row and the components of $\vec{\mathbf{A}}$ and $\vec{\mathbf{B}}$ in the second and third rows, respectively:

$$\vec{A} \times \vec{B} = \begin{vmatrix} \hat{i} & \hat{j} & \hat{k} \\ 2 & -1 & 1 \\ 3 & 4 & -3 \end{vmatrix} = \hat{i} \begin{vmatrix} -1 & 1 \\ 4 & -3 \end{vmatrix} - \hat{j} \begin{vmatrix} 2 & 1 \\ 3 & -3 \end{vmatrix} + \hat{k} \begin{vmatrix} 2 & -1 \\ 3 & 4 \end{vmatrix}$$

$$\vec{A} \times \vec{B} = \hat{i}(3-4) - \hat{j}(-6-3) + \hat{k}(8+3)$$
$$\vec{A} \times \vec{B} = -\hat{i} + 9\hat{j} + 11\hat{k}$$

In physics, the cross product is used, for example, in: the equation for torque and the relationship between angular momentum and linear momentum, which are fundamental to rotation; the Lorentz force exerted on a moving charge (or current) in an electromagnetic field; and Stokes's theorem, central to Faraday's law, which is the basis for an electric generator, and Ampère's law, which describes, for example, the attractive force exerted between two parallel current-carrying conductors.

Like the dot product, the magnitude of the cross product can also be expressed trigonometrically; the difference is that $\|\vec{A} \times \vec{B}\| = AB\sin\theta$ involves the sine function, whereas $\vec{A} \cdot \vec{B} = AB\cos\theta$ involves the cosine function. Here, A and B are the magnitudes of $\vec{A}$ and $\vec{B}$ and θ is the angle between them. Notationally, the double lines $\|\vec{A}\|$ mean to take the magnitude of the vector $\vec{A}$; e.g. $A = \|\vec{A}\|$.

The cross product between vectors $\vec{A}$ and $\vec{B}$ depends on the angle θ between $\vec{A}$ and $\vec{B}$: $\|\vec{A} \times \vec{B}\| = AB\sin\theta$. Noting that $\sin 0° = 0$, $\sin 90° = 1$, and that $\sin\theta$ increases from 0 to 1 as θ increases from 0° to 90°, the scalar product is maximum when $\vec{A}$ and $\vec{B}$ are perpendicular and zero when they are parallel.

Since $\sin 0° = 0$, two vectors $\vec{A}$ and $\vec{B}$ are parallel (i.e. $\theta = 0°$) if $\vec{A} \times \vec{B} = 0$ is zero. For any vector $\vec{A}$, $\vec{A} \times \vec{A} = 0$.

Torque is the rotational analog for force – a push or a pull. A net force is what is needed to overcome an object's inertia – a natural tendency to maintain constant velocity – causing the object to accelerate (i.e. to change its velocity). Rigid bodies (solids that preserve their structure rather than fly apart or deform when rotated) have rotational inertia – a natural tendency to rotate with constant angular velocity – also known as moment of inertia. Torque is what is needed to overcome a rigid body's rotational inertia, causing its angular velocity to change. A force is needed to exert a torque, though a force does not necessarily exert a torque: Where the force is applied and the direction of the force are also important factors for torque. As a cross product, torque $\vec{\tau}$ (the Greek letter tau) is defined as $\vec{\tau} = \vec{r} \times \vec{F}$, where $\vec{F}$ is force and $\vec{r}$ is the shortest-distance vector from the axis of rotation to the point where $\vec{F}$ is applied. The magnitude of the torque is given by $\tau = rF\sin\theta$.

A door is a common rigid body that can rotate. The axis of rotation runs through the hinges parallel to one edge of the door. In order to shut a stationary open door, it is necessary to exert a torque. In order to better understand how torque is involved in shutting a door, it is instructive to try some foolish things. First, with a door ajar, grab both handles and pull along the width of the door, away from the hinges. Why doesn't the door shut? There is a real force, and the distance r is not zero because it extends from the hinges to the doorknobs – so why is the torque zero? The answer is that $\vec{r}$ and $\vec{F}$ are parallel – i.e. $\theta = 0°$ – and $\sin 0° = 0$; therefore, the torque is zero and the door does not shut. Try pushing the edge of the door in different directions. Pushing the edge of the door toward the hinges similarly causes zero torque (since $\sin 180° = 0$). Varying θ, it is easiest to shut the door when $\theta = 90°$ – i.e. when the force is perpendicular to the door. Now try pushing the center of the door instead of the edge and it is more difficult to shut the door. Trying to push on the hinges is ineffective since $r = 0$ in this case.

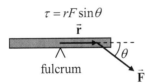

pulling a string tied to a nail on a rod causes the rod to rotate about the axis of rotation (at the fulcrum)

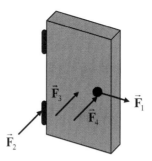

For $\vec{F}_1$, $\theta = 0°$. For $\vec{F}_2$, $r = 0$. $\vec{F}_4$ is more effective than $\vec{F}_3$ because r is greater.

The magnitude of the cross product $AB\sin\theta$ for two vectors $\vec{A}$ and $\vec{B}$ equals the area of the parallelogram defined by drawing $\vec{A}$ and $\vec{B}$ tip-to-tail. The reason is that the parallelogram can be split into two congruent triangles. The area of a triangle is one-half base times height. Taking A to be the base, $B\sin\theta$ is the height. Each triangle has area $AB\sin\theta/2$, so the parallelogram has area $AB\sin\theta$.

the magnitude of the cross product equals the area of the parallelogram formed by $\vec{A}$ and $\vec{B}$

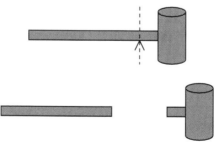

Puzzle 6.10: A croquet mallet is balanced on a fulcrum placed beneath its center of gravity. If the croquet mallet is cut into two pieces at its center of gravity, which piece will weigh more?

The vector algebra associated with the cross product has some interesting features compared to ordinary algebra. For one, the cross product is anti-commutative: $\vec{A} \times \vec{B} = -\vec{B} \times \vec{A}$. For another, the cross product is not associative, since $\vec{A} \times (\vec{B} \times \vec{C})$ is, in general, not equal to $(\vec{A} \times \vec{B}) \times \vec{C}$. However, the cross product is distributive with respect to vector addition, i.e. $\vec{A} \times (\vec{B} + \vec{C}) = \vec{A} \times \vec{B} + \vec{A} \times \vec{C}$.

Puzzle 6.11: Evaluate $\hat{i} \cdot \hat{i}$, $\hat{j} \cdot \hat{k}$, $\hat{k} \cdot \hat{j}$, $\hat{k} \times \hat{k}$, $\hat{j} \times \hat{k}$, and $\hat{k} \times \hat{j}$.

The cross product does not generalize to higher (or lower) dimensions. The reason is that the vector algebra of the cross product, which must be mathematically sound and generalize the physics formulas associated with the cross product in a plausible way, is well-known to only be valid in 3D and 8D, and is only physically meaningful in 3D. Aside from the mathematics, there is an obvious conceptual problem with generalizing the right-hand rule, which relates the direction of the vectors in the cross product, to higher-dimensions. Knowing this, it would be absurd to consider any theory where particles can propagate into extra dimensions if fundamental physics concepts such as angular momentum, torque, and magnetic field – involved in the cross product – were to make any sense since the cross product is not physically meaningful in extra dimensions; except for one important point that saves extra-dimensional theories from the pitfalls of the cross product: There is an alternative construction to the cross product, using matrix multiplication, that naturally generalizes to higher dimensions and is both mathematically sound and physically meaningful in any number of dimensions.

As it turns out, any cross product between two vectors in 3D can be expressed using one square matrix and two vectors; while the cross product only works in 3D, the matrix multiplication works in a space of any dimensionality. The three components that traditionally form a vector in 3D are instead interpreted to form the off-diagonal elements of an anti-symmetric square matrix.

6.6 Higher-Dimensional Matrices

The product of an $L \times M$ matrix **A** and an $M \times N$ matrix **B** is an $L \times N$ matrix $\mathbf{C} = \mathbf{AB}$. The (1,1) element of **C** is formed by matching the first column of **B** with the first row of **A**, multiplying the paired elements together and summing the products – thinking of the first column of **B** as a vector and the first row of **A** as another vector, the (1,1) element of **C** is the dot product between these column and row vectors. The (i, j) element of **C** is found by similarly pairing the jth column of **B** with the ith row of **A**.

$$\mathbf{A} = \begin{pmatrix} A_{11} & A_{12} & A_{13} \\ A_{21} & A_{22} & A_{23} \end{pmatrix} \qquad \mathbf{B} = \begin{pmatrix} B_{11} & B_{12} & B_{13} & B_{14} \\ B_{21} & B_{22} & B_{23} & B_{24} \\ B_{31} & B_{32} & B_{33} & B_{34} \end{pmatrix}$$

$$\mathbf{C} = \mathbf{AB} = \begin{pmatrix} A_{11} & A_{12} & A_{13} \\ A_{21} & A_{22} & A_{23} \end{pmatrix} \begin{pmatrix} B_{11} & B_{12} & B_{13} & B_{14} \\ B_{21} & B_{22} & B_{23} & B_{24} \\ B_{31} & B_{32} & B_{33} & B_{34} \end{pmatrix}$$

$$\mathbf{C} = \begin{pmatrix} C_{11} & C_{12} & C_{13} & C_{14} \\ C_{21} & C_{22} & C_{23} & C_{24} \end{pmatrix}$$

$C_{11} = A_{11}B_{11} + A_{12}B_{21} + A_{13}B_{31}$ $\qquad C_{21} = A_{21}B_{11} + A_{22}B_{21} + A_{23}B_{31}$
$C_{12} = A_{11}B_{12} + A_{12}B_{22} + A_{13}B_{32}$ $\qquad C_{22} = A_{21}B_{12} + A_{22}B_{22} + A_{23}B_{32}$
$C_{13} = A_{11}B_{13} + A_{12}B_{23} + A_{13}B_{33}$ $\qquad C_{23} = A_{21}B_{13} + A_{22}B_{23} + A_{23}B_{33}$
$C_{14} = A_{11}B_{14} + A_{12}B_{24} + A_{13}B_{34}$ $\qquad C_{24} = A_{21}B_{14} + A_{22}B_{24} + A_{23}B_{34}$

Example 6.3: Multiply the two square matrices below to form $\mathbf{C} = \mathbf{AB}$:

$$\mathbf{A} = \begin{pmatrix} 1 & 2 \\ 3 & 4 \end{pmatrix} \qquad \mathbf{B} = \begin{pmatrix} 0 & 1 \\ 2 & 3 \end{pmatrix}$$

Solution:

$$\mathbf{C} = \mathbf{AB} = \begin{pmatrix} 1 \cdot 0 + 2 \cdot 2 & 1 \cdot 1 + 2 \cdot 3 \\ 3 \cdot 0 + 4 \cdot 2 & 3 \cdot 1 + 4 \cdot 3 \end{pmatrix} = \begin{pmatrix} 4 & 7 \\ 8 & 15 \end{pmatrix}$$

Matrix algebra does not follow the same rules as ordinary algebra. For example, matrices do not, in general, commute – i.e. **AB** ≠ **BA**, except for a limited number of special cases – and matrix multiplication is not associative – i.e. **A**(**BC**) ≠ (**AB**)**C**, in general.

However, matrix multiplication is distributive with respect to matrix addition – i.e. $\mathbf{A}(\mathbf{B}+\mathbf{C}) = \mathbf{AB}+\mathbf{AC}$.

A 1×1 matrix has a single element, like a scalar. An $N\times 1$ or $1\times N$ matrix has N components, and may thus be regarded as a matrix representation of an N-dimensional vector. An $N\times 1$ matrix is a column vector, while a $1\times N$ matrix is a row vector. Notation: In matrix form, a column vector $|\vec{\mathbf{C}}\rangle$ is distinguished from a row vector $\langle\vec{\mathbf{R}}|$ by using bra $\langle\ |$ and ket $|\ \rangle$ notation.

$$|\vec{\mathbf{C}}\rangle = \begin{pmatrix} C_x \\ C_y \\ C_z \end{pmatrix} \qquad \langle\vec{\mathbf{R}}| = (R_x, R_y, R_z)$$

The usual dot product between vectors, referred to as the *inner product* in the context of matrices, results from multiplying a row vector $\langle\vec{\mathbf{R}}|$ by a column vector $|\vec{\mathbf{C}}\rangle$. Combining a bra $\langle\ |$ and ket $|\ \rangle$ in this fashion results in a bra(c)ket $\langle\ |\ \rangle$.

$$\langle\vec{\mathbf{R}}|\vec{\mathbf{C}}\rangle = (R_x\ R_y\ R_z)\begin{pmatrix} C_x \\ C_y \\ C_z \end{pmatrix} = R_x C_x + R_y C_y + R_z C_z = \vec{\mathbf{R}}\cdot\vec{\mathbf{C}}$$

A third, less common, way to multiply two vectors is to multiply a column vector $|\vec{\mathbf{C}}\rangle$ by a row vector $\langle\vec{\mathbf{R}}|$. This outer product results in an $N\times N$ matrix. The outer product is a very important part of forming complete sets in group theory.

$$|\vec{\mathbf{C}}\rangle\langle\vec{\mathbf{R}}| = \begin{pmatrix} C_x \\ C_y \\ C_z \end{pmatrix}(R_x\ R_y\ R_z) = \begin{pmatrix} C_x R_x & C_x R_y & C_x R_z \\ C_y R_x & C_y R_y & C_y R_z \\ C_z R_x & C_z R_y & C_z R_z \end{pmatrix}$$

A square (i.e. $N\times N$) matrix $\mathbf{S}$ is symmetric about the long diagonal (the diagonal elements have identical row and column indices, S_{ii}) if $S_{ij} = S_{ji}$. For a square anti-symmetric matrix $\mathbf{A}$, $A_{ij} = -A_{ji}$. An N-dimensional symmetric matrix has $\dfrac{N(N+1)}{2}$ independent components, while an N-dimensional anti-symmetric matrix has $\frac{N(N-1)}{2}$ independent components. Notice that $\frac{N(N+1)}{2} + \frac{N(N-1)}{2} = N^2$. For example, in 3D, a symmetric matrix has 6 independent components, while an anti-symmetric matrix has 3 independent components (a 3D square matrix has 9 components all together). The diagonal elements of an anti-symmetric must be zero because otherwise A_{ii} could not also equal $-A_{ii}$ as demanded by $A_{ij} = -A_{ji}$.

$$\mathbf{S} = \begin{pmatrix} S_{11} & S_{12} & S_{13} \\ S_{12} & S_{22} & S_{23} \\ S_{13} & S_{23} & S_{33} \end{pmatrix} \qquad \mathbf{A} = \begin{pmatrix} 0 & A_{12} & A_{13} \\ -A_{12} & 0 & A_{23} \\ -A_{13} & -A_{23} & 0 \end{pmatrix}$$

Observe that a 3D anti-symmetric matrix has the same number (three) of independent components as a 3D vector. This is not true, however, in lower or higher dimensions. As a result of this coincidence, physical quantities that are really anti-symmetric matrices may easily be confused as vectors in 3D: In 3D, if a physical quantity has 3 measurable independent components, it would naturally be interpreted as a vector. In fact, there are some vector quantities in the known universe – including torque, angular momentum, angular velocity, and magnetic field – that behave somewhat differently from other vectors; these pseudovectors are all involved in cross products.

An alternative to the cross product that does generalize to higher dimensions can be constructed by replacing the pseudovectors with anti-symmetric matrices. In this way, the cross product takes on one of two forms, depending on which vector in the usual cross product is replaced by an anti-symmetric matrix. Each form is a basic matrix multiplication, arguably a more natural construction than the artificial determinant construction that defines the cross product. Comparison with the cross product shows that this matrix multiplication is equivalent to the cross product in 3D.

$$\mathbf{C} = |\vec{\mathbf{A}}\rangle\langle\vec{\mathbf{B}}| - |\vec{\mathbf{B}}\rangle\langle\vec{\mathbf{A}}| = \begin{pmatrix} A_x \\ A_y \\ A_z \end{pmatrix}(B_x, B_y, B_z) - \begin{pmatrix} A_x \\ A_y \\ A_z \end{pmatrix}(B_x, B_y, B_z)$$

$$\mathbf{C} = \begin{pmatrix} A_x B_x & A_x B_y & A_x B_z \\ A_y B_x & A_y B_y & A_y B_z \\ A_y B_x & A_z B_y & A_z B_z \end{pmatrix} - \begin{pmatrix} B_x A_x & B_x A_y & B_x A_z \\ B_y A_x & B_y A_y & B_y A_z \\ B_y A_x & B_z A_y & B_z A_z \end{pmatrix}$$

$$\mathbf{C} = \begin{pmatrix} 0 & A_x B_y - B_x A_y & A_x B_z - B_x A_z \\ B_x A_y - A_x B_y & 0 & A_y B_z - B_y A_z \\ B_x A_z - A_x B_z & B_y A_z - A_y B_z & 0 \end{pmatrix}$$

$$\mathbf{C} = \begin{pmatrix} 0 & C_z & -C_y \\ -C_z & 0 & C_x \\ C_y & -C_x & 0 \end{pmatrix}$$

$$|\vec{\mathbf{C}}\rangle = -\mathbf{A}|\vec{\mathbf{B}}\rangle = -\begin{pmatrix} 0 & A_z & -A_y \\ -A_z & 0 & A_x \\ A_y & -A_x & 0 \end{pmatrix}\begin{pmatrix} B_x \\ B_y \\ B_z \end{pmatrix}$$

$$|\vec{\mathbf{C}}\rangle = -\begin{pmatrix} A_z B_y - A_y B_z \\ A_x B_z - A_z B_x \\ A_y B_x - A_x B_y \end{pmatrix} = \begin{pmatrix} C_x \\ C_y \\ C_z \end{pmatrix}$$

7 Force Fields

In this chapter, we will see that the fundamental forces of nature are observed to obey very geometric laws. With extra dimensions motivated by superstring theory as a theory of everything, we expect this geometric connection to persist in the case of extra dimensions. As it turns out, in order to preserve the geometric nature of the force laws, the mathematical form of the force laws must change somewhat depending upon the number of extra dimensions – and their nature (i.e. whether or not they are compactified in some way). We will explore these concepts in this chapter, where we focus on some of the direct physical ramifications of one or more extra dimensions. But first we shall begin by reviewing the fundamental forces for the benefit of readers who are not already quite familiar with them.

7.0 Gravitational Forces

In the known universe, any two massive objects exert an attractive force on one another according to Newton's law of universal gravitation, which is directly proportional to each of the masses m_1 and m_2 and inversely proportional to the square of their center-to-center separation r:

$$F = G \frac{m_1 m_2}{r^2}$$

The proportionality constant G, called the *gravitational constant*, is a universal constant equal to $6.67 \times 10^{-11} \text{ N} \cdot \text{m}^2/\text{kg}^2$.

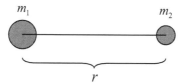

any two objects with mass attract each other gravitationally

Mass plays two fundamental roles in physics. One role is that mass is a measure of inertia – i.e. the reluctance of an object to be accelerated. It is much more difficult to accelerate (i.e. to change the speed of) a television, for example, than it is to accelerate a toaster because a television has much more mass. The other role that mass plays is that mass is the source of a gravitational field. Every mass creates an attractive gravitational field around it, and any two masses attract one another gravitationally.

The weight of an object is the force that a planet exerts on it. The weight W (not to be confused with work, which uses the same symbol) of an object is related to its mass m by the gravitational acceleration of the planet g_p: $W = mg_p$. The mass of an object does not depend on its position, but its weight does. For example, an object weighs less atop a mountain than it does near sea level.

> **Mass is a measure of an object's inertia – reluctance to accelerate – whereas an object's weight is the gravitational pull it experiences. An object's mass would be the same anywhere, but its weight depends upon its location – e.g. near the surface of the earth, on top of a mountain, on the moon, etc.**

According to Newton's law of universal gravitation, the force that a planet of mass m_p with radius R_p exerts on an object of mass m on the surface of the planet is

$$F = G\frac{m_p m}{R_p^2}$$

(assuming the object is small compared to the planet). Equating this force to the weight, mg_p, yields an expression for gravitational acceleration near the surface of a planet:

$$g_p = G\frac{m_p}{R_p^2}$$

For example, plugging in the mass and radius of earth yields a value for the acceleration of gravity near earth's surface.

The lowercase g represents gravitational acceleration, which describes how the velocity of an object in free fall changes in time. The uppercase G is the constant of proportionality in Newton's law of universal gravitation. Whereas G is a universal constant – i.e. the same value throughout the universe – g varies from place to place. These two "gee's" are fundamentally different, and even have different dimensions. The SI units of g are m/s^2, while the SI units of G are $\dfrac{m^3}{kg \cdot s^2}$.

An object is freely falling if the only force acting on it is the gravitational pull of a planet – i.e. its weight. In this case, its acceleration is gravitational acceleration g. Freely falling may be a little misleading in the sense that the object can actually be traveling upward. Throw an object upward and it continues to move upward until running out of speed, and then returns downward. Throw an object at an angle and it freely falls along a curved trajectory.

In addition to the law of universal gravitation, Newton has three fundamental laws of motion:

I. All objects have a natural tendency to maintain constant momentum called *inertia*.

II. The net external force acting on a system of objects equals the instantaneous time rate of change of the momentum of the system.

III. If Object 1 exerts a force on Object 2, then Object 2 simultaneously exerts a force on Object 1 that is equal in magnitude, but opposite in direction.

The momentum of an object is the product of its mass and velocity.

Inertia is the natural tendency of an object to maintain constant momentum, where momentum is mass times velocity (and velocity includes both speed and direction).

For a single object with constant mass (so, not a rocket ejecting steam, for example), constant momentum implies constant velocity. In this case, inertia is a natural tendency to maintain constant velocity (both speed and direction), and a net external force equals the product of the mass and acceleration of an object. Put another way, inertia is an object's resistance to accelerate. An object at rest wants to remain at rest, and a moving object wants to continue moving with constant velocity.

All forces inherently come in pairs. Any action comes with an equal, but opposite, reaction. The action is the force that Object 1 exerts on Object 2, and the reaction is the equal, but opposite, force that Object 2 exerts on Object 1. This is why recoil is experienced when firing a gun: The force that the gun exerts on the bullet is returned by an equal, but opposite, force that the bullet exerts on the gun (and shooter). Although the forces are equal in magnitude, the acceleration of the bullet (in the barrel of the gun) is much greater than the acceleration of the gun (and shooter) – i.e. the recoil – because the bullet has much less mass (and hence inertia).

Puzzle 7.1: A 240-lb. football player running 8 mph collides head-on with a 120-lb. cheerleader initially at rest. How much more force does the football player impart to the cheerleader than he receives in return?

The acceleration of an object in free fall is gravitational acceleration g, regardless of the mass of the object. The reason is that in free fall the net force acting on the object is $m\vec{g}$. According to Newton's 2nd law, the net force equals $m\vec{a}$. Equating these, mass cancels, and $\vec{a} = \vec{g}$. The gravitational mass in $m\vec{g}$ is exactly the same inertial mass in $m\vec{a}$ – i.e. the dual role of mass is why gravitational acceleration is independent of mass.

> **Acceleration is the instantaneous time rate of chance of velocity; conceptually, it describes how velocity (both speed and direction) changes. Gravitational acceleration is the acceleration an object would have if freely falling (i.e. the only force acting on the object is gravitational pull).**

A vacuum is a region of space completely devoid of matter (note that there can still be, and often is, gravity in a vacuum). Intergalactic space, for example, is a very near vacuum.

Puzzle 7.2: A 1-oz. feather and a 1-lb. rock are simultaneously released from rest from a 1-m height in a perfect vacuum chamber near earth's surface. How much sooner does the rock strike the ground than the feather?

For variations of altitude that are small compared to the radius of a planet, gravitational acceleration is approximately uniform – i.e. constant. For example, gravitational acceleration is 9.8 m/s^2 near the surface of the earth. This means that, neglecting air resistance, an object thrown straight upward loses about 10 m/s of speed each second and gains about 10 m/s of speed each second on the way back down.

Puzzle 7.3: In a train that is uniformly decelerating while heading north along horizontal rails, a dart is released from rest (relative to the train) directly above the bull's-eye of a target lying on the floor of the railway car. Relative to the bull's-eye, where does the dart land?

Puzzle 7.4: A rock is thrown straight upward near the surface of the earth. Neglecting air resistance, what is its acceleration at the top of its trajectory?

If an object is thrown horizontally or at an angle – i.e. not straight upward or downward – it follows a curved path. For a short change in altitude, as is the case with throwing a rock, the horizontal component of velocity is constant since no forces act horizontally, while vertically there is uniform acceleration from gravity pulling downward. The result is a parabolic path. Accounting for the variation of gravity with altitude, the path is actually an ellipse. The path is further modified from other effects, such as air resistance and the rotation of the earth.

Puzzle 7.5: Simultaneously, near earth's surface, one rock is released from rest while another is launched horizontally with a slingshot with a speed of 5 m/s – both from a height of 2 m. Which rock strikes the ground first?

A satellite is an object that orbits another planet, which is commonly a planet or star. There are numerous man-made satellites orbiting the earth, while moons and planets are celestial satellites. A satellite is really no different from a projectile except that it has enough tangential speed to stay in orbit. Launch a rocket with too little speed and it will crash into the earth, launch it with enough speed and it may wind up in orbit around the earth, and with greater speed yet it is possible to escape earth's pull all together.

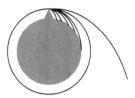

a satellite is a projectile with enough tangential speed to remain in orbit

Gravity is universal in the sense that the same law – namely, Newton's law of universal gravitation – explains terrestrial gravity – e.g. the motion of a rock thrown near the surface of earth – as well as celestial motion – e.g. the motion of the moons, planets, and sun in the solar system. Newton's law of universal gravitation evidently applies throughout the known universe, with the same form and same value for the gravitational constant.

Puzzle 7.6: The earth attracts the moon, exerting a gravitational force on it. Why doesn't the moon crash into the earth?

Kepler postulated three laws that describe behavior in the solar system:
I. Planets travel along elliptical orbits with the sun lying at one focus; moons follow similar orbits around planets.
II. A moon, planet, comet, meteor, asteroid, etc. sweeps out equal areas in equal times as it travels along its orbit.
III. The square of the orbital period of a planet or moon is proportional to the cube of the semi-major axis of its elliptical orbit.
Kepler's laws follow from other fundamental laws in physics. The possible orbits, for which Kepler postulated only the ellipse, can be derived from Newton's law of universal gravitation. That satellites sweep out equal areas in equal times is a consequence of conservation of angular momentum. Kepler's 3rd law also follows from Newton's law of universal gravitation.

Newton's law of universal gravitation leads to five possible orbits for two interacting massive objects: a straight line, a circle, an ellipse, a hyperbola, and a parabola, depending on the initial conditions – namely, the location of the two objects and their initial velocities. Released from rest, like an apple falling from a tree, yields a linear motion toward the center of mass of the system. The circle and ellipse are the two bound orbits for which the initial velocity is less than the escape velocity, and the parabola and hyperbola are the open orbits for which one object has enough kinetic energy (one-half mass times speed squared) to escape the gravitational pull of the other. There are a couple of improvements to Kepler's 1st law: For a sun and a single planet, it is actually the center of mass of the system (not the sun), that lies at one focus, but the sun has so much more mass than the other planets that the difference is slight; considering more than two objects, the orbits are not perfect conic sections – e.g. the moon wobbles around the earth considering the pull of the sun, Jupiter, and other celestial bodies.

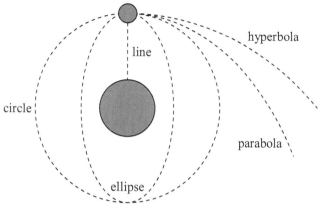

the allowed orbits in the two-body problem are conic sections

Angular momentum is the rotational analog for linear momentum. According to Newton's 1st law, objects have a natural tendency to maintain constant linear momentum. Similarly, rigid bodies – object that do not distort or break apart during rotation – have a natural tendency to maintain constant angular momentum. Linear momentum is mass times velocity, and angular momentum is moment of inertia times angular velocity. The moment of inertia of a rigid body depends on the mass of the rigid body as well as how the mass is distributed. More mass further from the axis of rotation leads to a greater moment of inertia.

According to Newton's laws of motion, linear momentum is conserved – i.e. the total linear momentum of a system of objects remains constant – if the net external force acting upon the system is zero. A net external force is needed to cause a change in linear momentum. Angular momentum is conserved when the net external torque is zero; a net external torque is required to change the angular momentum of a system of rigid bodies. Consider an ice skater spinning with her arms extended. When she brings her arms inward, she is bringing more of her mass closer to the axis of rotation, which decreases her moment of inertia. In order to conserve her angular momentum, her angular speed increases. Satellites similarly conserve angular momentum. Treating a satellite as pointlike, its angular momentum is $m\upsilon_T r$ (mass times transverse velocity times orbital distance). As a comet approaches the sun, its orbital distance decreases so its speed increases to compensate, conserving angular momentum. Similarly, as a planet orbits the sun along an ellipse, its speed is greatest at perihelion (nearest point to the sun) and least at aphelion (furthest point from the sun).

conservation of angular momentum for a spinning ice skater

The orbital period of a satellite is the time it takes to complete one orbit. Kepler noted that the square of the orbital period of a planet is proportional to the cube of the semi-major axis of its elliptical orbit.

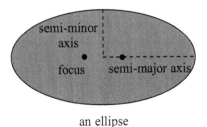

an ellipse

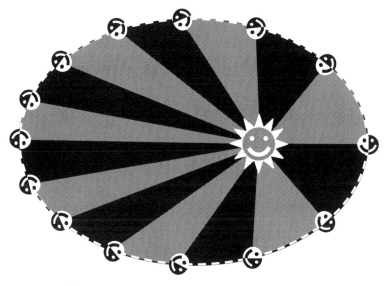

a planet sweeps out equal areas in equal time intervals

Astronauts in a spaceship in a free fall orbit experience weightlessness, as if in zero-gravity. There is, in fact, gravity, and the astronauts do have weight, although not as much as near the surface of earth. Rather, the spaceship and astronauts fall freely, so the astronauts do not experience a support force.

Scales designed to measure the weight of an object actually measure what is called *apparent weight*. The apparent weight measured by a bathroom scale, for example, equals the support force that the scale exerts on the person standing on it. When the bathroom scale is not accelerating, this measured support force equals the weight of the person. However, if the bathroom scale is used in an elevator that is accelerating upward, the scale will read a force in excess of the actual weight of the person since the net force acting on the person equals the person's mass times acceleration according to Newton's 2nd law. The apparent weight read by the scale in this case exceeds the actual weight of the person. If the elevator instead accelerates downward, the apparent weight is less than the actual weight. In the unfortunate event that the cable snaps, the scale will read zero. There is no upward support force in this free fall orbit. It is not just that the scale reads zero, but the person actual feels weightless. A person does not feel the force of gravity pulling downward; rather, what is felt is the force of the ground, or a chair, or something exerting an upward

45

support force to prevent a downward fall. While jumping in the air, a person feels temporarily weightless (except for the force of air resistance). Jumping off a tall building, the force of the fall is not felt at all; it's when the fall ends that a great force is suddenly experienced.

Puzzle 7.7: A scale reads 150-lbs. when a passenger stands on it in an elevator car that is at rest. What does the scale read when the elevator car travels upward at a constant speed of 5 m/s?

Owing to the smallness of the gravitational constant G in SI units, 6.67×10^{-11} N·m^2/kg^2, Newton's law of universal gravitation is most significant if at least one of the masses is astronomical. Two semi trucks with a mass of 1,000 kg each separated by a couple of meters exert a gravitational force of $\sim 10^{-5}$ N. For two objects separated by about a meter, the product of their masses must be about 10^{11} kg^2 in order to attract one another with a measly 1-N force.

Puzzle 7.8: There are two ways to obtain more gravitational force. One way is to increase the mass of one (or both) objects. The other way is the make the separation r smaller. Explain why it is not physically possible to make the force between two semi trucks large by making r smaller. In fact, it is not possible for r to be as small as 1 meter in the case of two semi trucks! What's the catch?

7.1 Electromagnetic Forces

In the known universe, any two particles with electric charge exert a force on one another according to Coulomb's law, which is directly proportional to the charges q_1 and q_2 and inversely proportional to the square of their separation r:

$$F = k \frac{q_1 q_2}{r^2}$$

The proportionality constant k, called *Coulomb's constant*, equals $9.0 \times 10^9 \, \text{C} \cdot \text{m}^2/\text{kg}^2$.

any two charged particles attract or repel each other electrically

Electric charge is the source of an electric field. Every charged particle creates an electric field around it, and any two charged particles interact electrically. Electric charge can be positive (like the proton) or negative (like the electron). Charges of the same sign repel, while charges of opposite sign attract. The SI unit of electric charge is the Coulomb (C).

Matter is composed of various types of atoms, which consist of protons, neutrons, and electrons. A proton has a charge of 1.6×10^{-19} C, a neutron is electrically neutral, and an electron has a charge of -1.6×10^{-19} C. Macroscopic objects can have zero net charge if the number of protons balances the number of electrons.

Coulomb's law has the same mathematical form as Newton's law of universal gravitation. However, there are some major differences between these two laws:
- For one, electric charges come in two varieties – positive and negative – whereas mass does not.
- Similarly, the gravitational force between two masses is always attractive, but the force between charged particles can be attractive or repulsive.
- Two charged particles of the same sign repel, but two masses attract (and, of course, the two masses are both positive).
- There exist neutral particles (such the neutron), which have no electric charge. There do exist particles with zero rest-mass (like the photon), but even the photon has relativistic (or inertial) mass.
- Macroscopic objects cannot have zero mass, but most macroscopic objects are electrically neutral (i.e. zero net charge) because they often have the same number of protons as electrons.
- Mass plays two fundamental roles – mass is a measure of inertia, and the source of a gravitational field. Electric charge is only the source of an electric field, not a measure of inertia. The electron is more easily accelerated than the proton since, although their

charge is equal and opposite, the electron has much less rest-mass.
- The proportionality constant for gravity, 6.67×10^{-11} Nm^2/kg^2, is very small in SI units, while the proportionality constant in Coulomb's law, 9.0×10^9 Cm^2/kg^2, is $\sim 10^{20}$ times greater in magnitude in its SI units. Combined with the fact that elementary charge, 1.6×10^{-19} C (the charge of one proton), is much greater in magnitude in SI units than the masses of elementary particles (e.g. the mass of the proton is 1.67×10^{-27} kg), Coulomb's law is very significant at the microscopic scale whereas gravity is, generally, comparatively insignificant (by a factor on the order of 10^{-36} to 10^{-42} at the microscopic scale) except for astronomical masses.
- Moving electric charges create not only electric fields, but also magnetic fields. Moving electric charges constitute electric current. Electric current is the source of a magnetic field.

Imagine a universe very much like the known universe, except that it has negative masses. Assume that the laws of physics of this hypothetical universe are the same as those of the known universe. Furthermore, take positive masses to behave the same way as all masses behave in the known universe. Then, like masses would attract and opposite masses would repel (different from the behavior of charge). Living on a planet of positive mass would be very much the same as living on earth in the known universe. It would be very unlikely for any objects with negative mass to be present, since they would be repelled by the tremendous positive mass of the planet. If a rock with negative mass were dug from a deep hole, for example, the rock would fly upward, gaining speed as it leaves the gravitational push of the earth. All of the negative mass of this hypothetical universe would tend to collect on one side, as far as possible from the positive mass on the other side. Considering this, it seems plausible that there could be such objects with negative mass in the known universe, since they could all exist so far away from the observable universe that they remain undetected. However, no such negative masses have ever been produced in high-energy colliders, and cosmological models dating back to the Big Bang do not appear to suggest a possible division of the universe into two repelling mass states.

Electric charge is quantized – i.e. the net electric charge on any object is an integer times the unit of elementary charge, $e = 1.6 \times 10^{-19}$ C. The reason for this is obvious for ordinary, macroscopic matter observed in the universe: Since such matter consists of protons, neutrons, and electrons, the net electric charge is proportional to the number of protons minus the number of electrons. A chunk of matter that has 8 more electrons that protons, for example, has a net charge of $-8e$. (Notice that e is a positive value corresponding to the electric charge of a proton such that $-e$ is the electric charge of an electron.) The quantization of electric charge states that it is impossible to find an object in nature with an electric charge of $1.5e$, which would be like having one and a half protons. There are other particles in the universe besides protons, electrons, and neutrons, but the electric charge of any object observed in the universe is always an integral multiple of e.

Protons and neutrons are not actually elementary particles: As it turns out, protons and neutrons are composed of other particles called quarks. The proton consists of two up quarks and a down quark, and the neutron consists of two down quarks and an up quark. The proton and neutron have nearly equal mass, but the neutron is slightly greater. (The fact that the neutron is slightly greater corresponds to the fact that free neutrons can decay, but protons are observed to be stable.) The down quark has slightly more rest-mass than an up-quark. The down quark has an electric charge of $-e/3$, and the up quark has an electric charge of $2e/3$. This gives the proton an electric charge of e and the neutron zero electric charge. Although the quarks have fractional electric charge, all observed states in the universe still have integral electric charge. The reason is that individual quarks cannot exist freely: Rather, quarks only exist in nature in combinations for which the net electric charge is an integral multiple of e.

Magnets have apparent north and south poles, which behave like electric charges in the sense that opposite poles attract and like poles repel. For example, the north pole of one magnet is attracted to the south pole of another magnet, but repelled by the other north pole.

The pair of opposite poles – i.e. north and south – present in a magnet form what is called a *magnetic dipole*. The north and south poles always come together: It is not possible to isolate just a north or south pole. If it were possible to isolate a single pole, it would be referred to as a magnetic monopole; however, magnetic monopoles have never been detected. A naïve attempt to isolate one pole of a magnet is to cut the magnet in half. Instead of two half magnets each containing one pole, the original magnet splits into two smaller magnets, each with its own north and south pole. Cutting one of these smaller magnets in half yields the same result. North and south poles always exist together in pairs, never alone. Cutting a magnet in half successively, eventually all that will be left is a single atom.

Electric monopoles, on the other hand, do exist. It is possible to have a single positive (or negative) electric charge. The electron, for example, is an electric monopole.

Moving electric charges produce magnetic fields. Two moving electric charges exert a magnetic force on one another in addition to the electrical attraction or repulsion described by Coulomb's law. The electrically charged particles (protons and electrons) inherently present in matter ultimately give rise to the net magnetic field of a magnet.

Mass is the source of a gravitational field, electric charge is the source of an electric field, and moving charge (or current) is the source of a magnetic field.

Technically, even a stationary particle with electric charge can produce a magnetic field – if it has a property called *spin*. A particle moving in a circle has orbital angular momentum. If such a particle has electric charge, its orbital angular momentum creates a magnetic field. However, most particles have an intrinsic spin angular momentum. That is, they behave as if they have angular momentum even if they are at rest. If such a particle has electric charge, its spin angular momentum creates a magnetic field – even if it is stationary. The name *spin* is associated with an analogy with earth. The earth has two kinds of angular momentum: It has orbital angular momentum associated with its revolution around the sun, and it has spin angular momentum in the sense that it spins about an axis. A point-particle, such as an electron, cannot really be thought of as spinning, though, so its spin angular momentum is said to be an intrinsic property.

Most particles have a property called spin angular momentum (or simply spin), for which they behave as if they have angular momentum even when they appear to be at rest.

The protons, electrons, and neutrons (remembering that neutrons are composed of quarks, which individually have electric charge even though the composite object is neutral) that make up a material individually contribute tiny magnetic fields. If these magnetic fields are randomly aligned, the material is nonmagnetic; but if these magnetic fields are at least partially aligned, the material acts as a magnet, behaving as if it had two well-defined poles.

the earth has orbital and spin angular momentum

The north pole of a compass needle points toward a point near the geographic north pole of the earth. Of course, the north pole of a magnet is attracted to a magnetic south pole. Thus, although the compass points near geographic north, it is really pointing to magnetic south. The earth behaves like a giant bar magnet with the magnetic poles approximately reversed compared to the geographic poles.

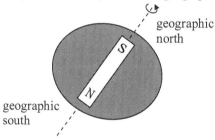

the earth has a magnetic field as if it were a giant bar magnet

An electric current is a stream of electrically charged particles. In a conducting wire, for example, the electric current is proportional to the number of electrons passing though a given point in the circuit per unit time (more precisely, it is the instantaneous time rate of change of electric charge). Since a moving particle with electric charge produces a magnetic field, an electric current can create a substantial magnetic field (with contributions from the magnetic fields of numerous individual charged particles).

Two long, parallel wires carrying electric currents I_1 and I_2 separated by a distance r (between their centers) in vacuum exert an attractive magnetic force on one another:

$$F = \mu_0 \frac{I_1 I_2 L}{2\pi r}$$

where L is the length of each wire (assumed to be the same) and the proportionality constant μ_0, called the permeability of free space, is $4\pi \times 10^{-7} \, \text{N}/\text{A}^2$. If the currents are instead antiparallel (i.e. in opposite directions), the force is repulsive instead of attractive.

One of the great surprises in the history of electricity and magnetism is that electricity and magnetism are two different manifestations of a more fundamental, underlying phenomenon. On the surface, electricity and magnetism seem to be completely unrelated. It is amazing that the same physics that ultimately explains why a charged rubber rod attracts a charged glass rod also explains why the north pole of compass needle points to a point near the geographic north pole of the earth. The charged rods attract because one rod has an excess of electrons (compared to the number of protons it has) while the other has a deficiency of electrons, and oppositely charged objects attract. The compass needle and earth are both magnets, whose magnetic fields arise due to the motion of electric charges. The fact that electricity and magnetism both relate to electric charges, however, does not itself unify the separate electrical and magnetic phenomena into a single electromagnetic phenomenon. Electricity and magnetism are inherently related through other experimental observations: A changing magnetic field can produce an electric field, and a changing electric field can produce a magnetic field. Furthermore, light is observed to be an electromagnetic wave, which has intimately related oscillating electric and magnetic fields – really, one oscillating electromagnetic field. All electrical and magnetic phenomena can be explained in terms of a single, unifying electromagnetic field.

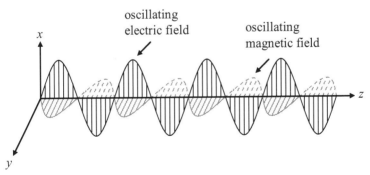

an electromagnetic wave propagating in the positive z-direction; the electric and magnetic fields oscillate perpendicular to the direction of propagation and to each other

Faraday discovered that a changing magnetic field induces an electric field. For example, a magnet moving through a coil of wire induces an electric current in the wire. The moving magnet produces a moving magnetic field. This changing magnetic field induces an electric field in the wire, which causes electric charges in the conducting wire to accelerate; this stream of electric charges is electric current in the wire. The beauty of this is that Faraday's law, as it is called, provides a means to generate electric current without using a conventional battery or power supply, but by using magnets. An electric generator uses Faraday's law: Electricity can be generated by rotating a coil of wire between the opposite poles of two magnets. The relative motion between the coil of wire and the magnetic field induces electric current. The fact that a changing magnetic field can induce an electric field shows that these two fields are inherently related.

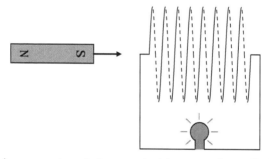

a moving magnet can induce an electric current in a coil of wire

7.2 Field Lines

Any spherically symmetric object with mass m produces an attractive gravitational field $\vec{g}$ that varies as the inverse-square of the distance from the center of mass of the object:[2]

$$g = \frac{Gm}{r^2}$$

---

[2] This is true at positions outside of the object; inside of the object, like the core of the earth, is a different matter. We will return to this point when we investigate Gauss's law.

The gravitational field of a system of objects can be visualized by drawing gravitational field lines (called *lines*, even though they often curve). The direction of the gravitational field lines is such that the gravitational field lines run toward massive objects. Although, technically, a gravitational field exists everywhere in space, it is visually useful to draw a diagram with a finite scale – such as drawing 1 field line per 1000 kg of mass. In this way, a finite number of gravitational field lines enter a given massive object, and the concentration of gravitational field lines on the diagram is proportional to the strength of the gravitational field at a given point.

The following information can be inferred from analyzing a sketch of the gravitational field lines for a system of objects:
- If an object were in a state of free fall at a given point on the sketch, the direction of the gravitational field at that point indicates in what direction the object would feel pushed.
- The gravitational field lines indicate the acceleration that an object would have in free fall as well as the weight of an object at a given point in the diagram. However, the gravitational field lines do not, in general, illustrate the trajectory of a freely falling object.
- Regions where gravitational field is relatively strong or weak can be inferred from the concentration of gravitational field lines. The greater the concentration, the stronger the gravitational field.
- The relative masses of different objects can be compared by counting the number of gravitational field lines intersecting each object. The more gravitational field lines intersecting an object, the greater its mass.

The gravitational field lines outside an isolated planet are radially inward (like the spokes of a bicycle wheel). This is because, at any point, the gravitational field is directed toward the center of mass of the planet. Another mass placed in the gravitational field of the planet would feel attracted along the gravitational field lines – i.e. toward the center of mass of the planet.

For a given system of massive objects, the gravitational field lines are related to a set of equipotential surfaces. An equipotential surface is a surface where the gravitational potential energy is constant (for a hypothetical *test mass*). Releasing a test mass from rest at a given point in the diagram, gravity would do work on the test mass, causing it to be displaced. Gravitational potential energy is a measure of how much work gravity can do in displacing an object. For an isolated massive object, the equipotential surfaces form a set of concentric spheres because no work is done transporting a test mass around the surface of any of these concentric spheres (all points on the surface of one of these spheres being no closer than any other to the isolated massive object in the center).

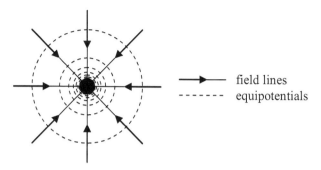

gravitational field lines for a planet

Note: The field maps shown here for 3D (and higher) spaces are 2D cross sections. In this diagram, for example, the field lines radiate inward three-dimensionally, and the equipotentials drawn here are really spherical surfaces (not circles).

The equipotential surfaces serve as a sort of topological map. The topological map of a mountain, for example, consists of curves of constant altitude spaced in even increments of altitude. The spacing between curves represents the change in altitude. The greater the spacing between two curves, the shallower the slope (on average) between them, and the smaller the spacing, the steeper the slope (on average). A mathematical operator called the gradient provides a measure of the steepness of the slope for a topological map. The gradient also provides a directional measure – outward (or inward) from one equipotential to the next. For a map of gravitational field lines, the equipotentials are surfaces – not curves (like a usual topological map for land) – and the gravitational field lines are proportional to the negative of the gradient of the gravitational potential energy of a hypothetical test mass.

Owing to mathematical properties of the gradient operator, the gravitational field lines always pass perpendicularly through the equipotential surfaces.

Volume 2: The Physics of the Fourth and Higher Dimensions...

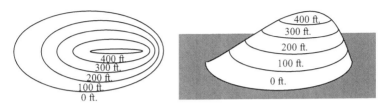

elevation contours form a topological map for an island

Puzzle 7.9: Can two equipotential surfaces intersect? Can two contour lines on a topological map intersect? Explain.

For a system of massive objects, the gravitational field lines are drawn according to the principle of superposition: The net gravitational field at any point is the vector addition of the gravitational fields produced by each massive object in the system. The net gravitational field is thus drawn by combining the individual gravitational fields tip-to-tail.

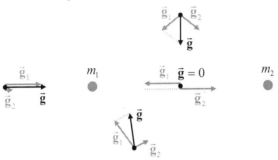

the superposition of gravitational fields ($m_1 = m_2$)

The gravitational field lines for a system of two massive objects, such as the earth-moon system, look radially inward near each massive object and on the far side of each object, but feature an interesting saddle point at the center of gravity of the system.

An object with a net electric charge produces electric field lines, which are similar to the gravitational field lines produced by a massive object except for one important difference: Electric field lines emerge from positive electric charges and converge at negative electric charges, whereas gravitational field lines only converge at massive objects.

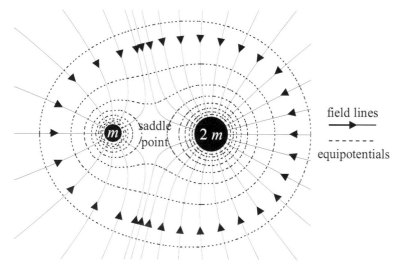

gravitational field map for a binary system

The electric field lines for a single, isolated positive charge, such as a proton, are radially outward, while for a single isolated negative charge, like an electron, they are radially inward.

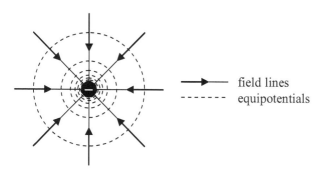

electric field map for an isolated negative charge

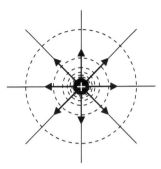

electric field map for an isolated positive charge

One field line diagram very fundamental to electricity, but not possible for gravity, corresponds to the electric dipole – i.e. a system with one positive and one negative electric charge. The field lines for the electric dipole are football-shaped between the electric charges and more radial near each electric charge or on the far side of each electric charge.

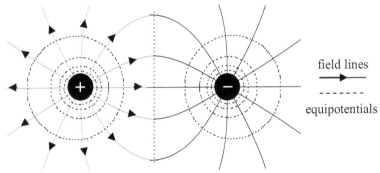

electric field map for the electric dipole

The magnetic field lines produced by a bar magnet closely resemble the electric field lines of the electric dipole, except for one important regard: The magnetic field lines behave much differently inside the magnet than the do outside the magnet. The resemblance outside the magnet is not surprising, since the poles of a magnet attract and repel like electric charges. The strange occurrence inside the magnet is related to the conceptual problem of trying to cut a magnet in half to isolate one of the poles. The magnetic field lines outside of the magnet behave as if there were two well-defined poles, but inside the magnet the magnetic field lines reveal that the fact that there are no magnetic monopoles, and the magnetic field of the magnet is really the superposition of numerous atomic magnetic fields.

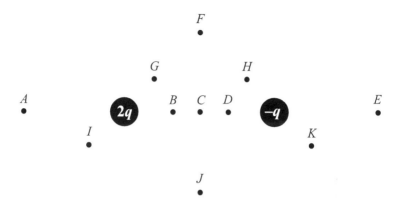

Puzzle 7.10: At which point(s) in the diagram above could the net electric field equal zero?

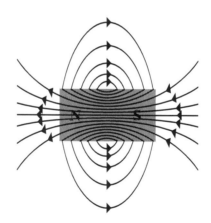

magnetic field lines for a bar magnet

Just as massive and charged objects produce gravitational and electric field lines, respectively, moving charges or currents produce magnetic field lines. However, the direction of the magnetic field lines produced by a moving charge or current is not nearly as intuitive. Whereas gravitational and electric field lines for an isolated point-source radiate inward or outward, the magnetic field lines for an isolated moving charge or a current travel in circles around the source.

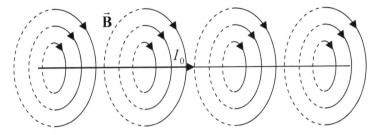

magnetic field lines for an infinite, straight wire

There is a useful right-hand rule for determining the direction of the magnetic field. With the thumb of your right hand in the opposed position, point your right thumb along the velocity of the moving charge (if positive – if negative, point them against the velocity) or along the direction of the current, as appropriate. Leaving the thumb opposed and pointing this way, curl the fingers of your right hand in a circular arc. This way, your thumb represents the velocity of the moving charge or the current and your curled fingers represent the circular magnetic field lines that surround the source. The magnetic field vector is tangential to these circles, and runs from your palm toward your fingertips.

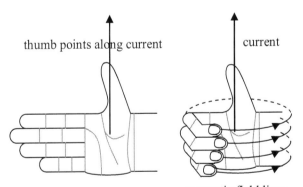

a right-hand rule for determining the direction of magnetic field lines: with the thumb pointing along the current, the curled fingers represent circular magnetic field lines

Puzzle 7.11: Draw the direction of the magnetic field at each point indicated in the diagram below (where I is the symbol for electric current).

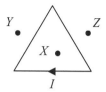

Puzzle 7.12: Draw the direction of the magnetic field at each point indicated in the diagram below for the electron that is moving upward.

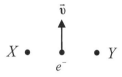

Puzzle 7.13: What direction of the current in the loop is needed to produce the magnetic field lines illustrated below (where $\vec{B}$ represents magnetic field)?

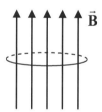

Puzzle 7.14: Draw the current that could produce the magnetic field lines illustrated below (where a circle with a dot in the center, ⊙, represents the tip of an arrow coming out of the page, while a circle with an X in the center, ⊗, represents the tail feathers of an arrow going into the page).

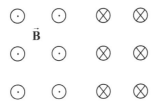

7.3 Gauss's Law

When analyzing a field line map, useful comparisons can be made by measuring the relative number of field lines at various points on the map. This can be achieved by visualizing imaginary surfaces at different locations in the diagram. The flux is a measure of the relative number of field lines passing through a surface. (The absolute number of field lines would be infinite since, in principle, there is a field line everywhere in the diagram. However, it is practical to adopt a scale – e.g. 5 field lines per $6\,\mu C$ of charge – which fixes a finite number of field lines in the diagram, so that the relative number of field lines can be compared.)

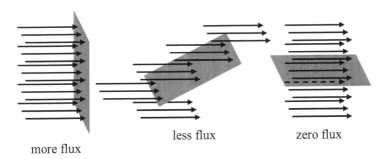

more flux   less flux   zero flux

    Raindrops offer a more tangible sense of flux. Consider the surface of an umbrella, shielding a pedestrian from falling rain. A flux of raindrops strikes the surface of the umbrella at any given moment. In the absence of wind, the rain falls straight down, and a small umbrella raised overhead effectively shields the pedestrian from the flux of raindrops. With strong winds blowing such that the rain falls at an angle, the flux of raindrops falling toward the pedestrian is increased (the pedestrian has a greater cross section in this case) and the same small umbrella does not offer full protection from the rain.

less flux   more flux

    Some of the field lines passing through a surface may be heading outward, while others may be heading inward. Outward field lines contribute toward a positive flux, while inward field lines contribute toward a negative flux. If the number of outward field lines equals the number of inward field lines, the net flux is zero. For a closed surface, the net flux is proportional to the number of outward field lines minus the number of inward field lines.

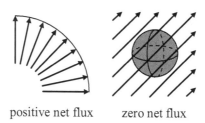

positive net flux   zero net flux

    According to Gauss's law, the net flux through any closed surface is proportional to the source enclosed by the surface. For gravity, the source is mass; for electricity, the source is electric charge; and for magnetism, the source is electric current. For an electric field map, for example, the net flux through any closed surface is proportional to the net electric charge inside the closed surface: If there is no electric charge enclosed or if there is just as much negative as positive electric charge enclosed, the number of outward electric field lines equals the number of inward electric field lines; if there is more positive than negative electric charge, there will be more outward than inward electric field lines; and if there is more negative than positive electric charge, the number of inward electric field lines will exceed the number of outward electric field lines.

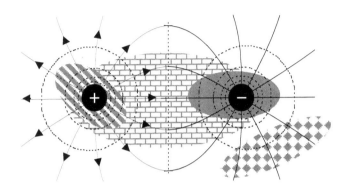

electric field map for the electric dipole

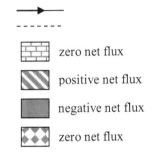

In electricity, for example, Gauss's law is very practical for calculating the electric field due to a distribution of electric charge for which there is a very high degree of symmetry. Gauss's law relates the net electric flux to the net charge enclosed by an arbitrary closed surface, called a Gaussian surface. Since the net electric flux is related to the electric field, Gauss's law provides an indirect, yet sometimes useful, means of calculating the electric field due to a distribution of electric charge. For electric charge distributions with enough symmetry, it is possible to choose a Gaussian surface for which it is conceptually simple to show that the electric field is constant throughout the surface. In such cases, the electric field is trivially related to the net electric flux so that Gauss's law provides a very simple calculation. Gauss's law is deceptive this way: It is expressed eloquently as a mathematical equation in vector calculus,

$\oint_S \vec{E} \cdot d\vec{A} = 4\pi k q_{enc}$ in electricity, yet it is especially useful as a conceptual tool. In gravity, the right-hand side becomes $4\pi G m_{enc}$ and the left-hand side has gravitational rather than electric flux.[3]

The flux through a surface is easiest to evaluate when the field – gravitational, electric, or magnetic – is constant in magnitude over the surface and the direction of the field is perpendicular or tangent to the surface at all points on the surface. When the field is constant and everywhere perpendicular to the surface, the flux is simply gA, EA, or BA, depending upon the nature of the field. When the field is tangent to the surface, the flux is zero (regardless of whether the field is constant in magnitude). Although Gauss's law holds for an arbitrary Gaussian surface, it is very useful to apply it to a Gaussian surface for which the flux is easy to calculate. The conceptual part of applying Gauss's law is choosing such a Gaussian surface and proving, through qualitative arguments, that the field is constant in magnitude and everywhere perpendicular to or tangent to the surface. This is much easier to prove when the source – mass or electric charge – has a highly symmetrical distribution.

Consider the following model for a planet, which illustrates a few important principles despite its simplicity: Imagine a perfectly spherical planet with mass m_p and radius R_p with uniformly distributed mass. The gravitational field of such a planet is spherically symmetric: It depends only upon the distance from the center of the planet – there is no preferred direction. The natural choice for a Gaussian surface outside the planet is a sphere of radius $r > R_p$ concentric with the planet: At every point on such a surface, gravitational field is constant in magnitude and points radially inward – perpendicular to the surface. Since the surface area of the Gaussian sphere is $4\pi r^2$, the net gravitational flux through this Gaussian surface is $gA = 4\pi g r^2$. All of the mass of the planet, m_p, is enclosed by the Gaussian surface. Thus, Gauss's law gives $4\pi g r^2 = 4\pi G m_p$, which simplifies to $g = G \dfrac{m_p}{r^2}$ for the gravitational field outside of a planet. Outside the planet, Gauss's law agrees with Newton's law of universal gravitation.

---

[3] Technically, there is also a directional minus sign.

Inside such a planet, the gravitational field is again spherically symmetric, suggesting a Gaussian sphere of radius $r < R_p$. The net gravitational flux is again $4\pi gr^2$. What is different inside the planet is that only some of the mass of the planet is enclosed by the Gaussian surface. The enclosed mass is proportional to the enclosed volume: $\frac{m_{enc}}{m_p} = \frac{V_{enc}}{V_p}$. Since the volume of a sphere is $\frac{4\pi r^3}{3}$, $\frac{V_{enc}}{V_p} = \frac{r^3}{R_p^3}$.

Solving for the mass enclosed by the Gaussian sphere, $m_{enc} = \frac{r^3}{R_p^3} m_p$.

Inside the planet, Gauss's law gives $4\pi gr^2 = 4\pi G \frac{r^3}{R_p^3} m_p$, which simplifies to $g = G \frac{m_p r}{R_p^3}$. Inside, the gravitational field is proportional to the distance from the center.

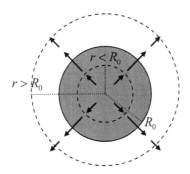

Gaussian surfaces inside and outside of a spherical source

For a sphere of radius R_0 with a uniform distribution of electric charge Q, Gauss's law yields an expression for the electric field that is analogous to the gravitational field of a planet: $E = k \frac{Q}{r^2}$ outside of the charged sphere and $E = k \frac{Qr}{R_0^3}$ inside the charged sphere. Notice that these two equations agree at the boundary $(r = R_0)$ – i.e. electric field is continuous across the boundary.

Gauss's law is not very practical for calculations in magnetism, but is still conceptually very significant. In fact, Gauss's law shows that the net magnetic flux is always zero. This follows from the observation that magnetic monopoles have never been discovered in the known universe. A magnet seems to behave as if it has two magnetic monopoles – a north pole and a south pole: Outside of the magnet, the magnetic field lines look like the electric field lines of an electric dipole. However, inside the magnet, the behavior is drastically different: While the field lines run from north to south outside the magnet, they reverse direction inside. This strange behavior inside the magnet is associated with the fact that if a bar magnet is broken, it divides into two smaller magnets. Not only is it impossible to break a magnet and isolate a north or south pole, no experiment has ever been successful in finding a magnetic monopole.

Although there are apparently no magnetic monopoles in the known universe, there are electric monopoles. Electric charges do not inherently come in pairs. Unlike a magnet, the electric charges of an electric dipole can be separated and isolated. A single electric charge constitutes an electric monopole.

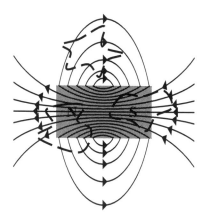

the net magnetic flux through any closed surface is always zero

According to Gauss's law, the net flux passing through any closed surface is proportional to the source (mass or charge) enclosed.

7.4 Stokes's Theorem

Gauss's law is a very geometric law involving an arbitrary closed surface and the net flux of field lines passing through that surface. Stokes's theorem is dimensionally analogous to Gauss's law in a sense. For one, Stokes's theorem considers a closed curve instead of a closed surface. Another difference is that Stokes's theorem involves the contribution of field lines that have components tangential to the closed curve, whereas the net flux in Gauss's law involves the contribution of field lines that have components perpendicular to the surface. These two laws are also related to properties of vectors: Gauss's law can be cast in a form where the field is involved in a dot product, while Stokes's theorem in similar form has the field involved in a cross product – two fundamental ways to multiply vectors. (More technically, Gauss's law in differential form involves the divergence of the field and Stokes's theorem in differential form involves the curl.)

The net flux through a surface can be thought of as an infinite sum, called a Riemann sum, as follows. Divide the surface up into tiny chunks. Determine the area of each chunk and the average value of the field over the chunk. For each chunk, calculate the field times the area times the cosine of the angle between them (where the area's direction is perpendicular to the surface, directed outward if the surface is curved). Sum the contribution from each chunk to obtain the net flux through the surface. This method serves as an approximation for the net flux. To obtain a better approximation, divide the surface up into a larger number of tinier chunks. In the limit that the chunks' areas become infinitesimal, this sum becomes infinite and the result is exact. This is called a Riemann sum, and the infinitesimal chunks are called differential elements. This probably seems to be a very inefficient and sloppy way to compute the net flux, but it is useful both conceptually and numerically. The Riemann sum serves as the conceptual basis of integration – an algebraic method of calculus for achieving the same result more directly (however, not every expression can be integrated algebraically, which is when the numerical prescription becomes very practical).

Stokes's theorem involves the line integral of the field along a closed curve instead of the net flux through a surface. The line integral can also be expressed as a Riemann sum. The closed curve is divided up into tiny chunks. For each chunk, determine its arc length and the average value of the field over the chunk and compute the average value of the field times the arc length times the cosine of the angle between them. The direction of the arc length is tangential to the curve, whereas the direction of area is perpendicular to a surface.

Stokes's theorem relates a line integral of a field along a closed curve to an integral over an open surface bounded by the closed curve. This is analogous to Gauss's law as Gauss's law relates an integral of a field over a closed surface (i.e. the net flux) to an integral over the volume enclosed by the surface (this integral determines the net charge enclosed by the surface). The analogy is geometric: A closed curve bounds an open surface, while a closed surface bounds a volume. Two complementary physical laws illustrate Stokes's theorem: Ampère's law and Faraday's law.

Ampère's law states that the line integral of magnetic field along any closed curve is proportional to the net electric current passing through any open surface bounded by the closed curve (assuming a steady current). This holds for any closed curve, called an Ampèrian loop. Similar to Gauss's law, Ampère's law is particularly useful when there is enough symmetry that it is easy to find an Ampèrian loop for which the magnetic field is constant in magnitude over the Ampèrian loop and always tangent or perpendicular to the Ampèrian loop. When the magnetic field is constant in magnitude over the Ampèrian loop and tangential to the loop, the line integral equals BC, where C is the circumference of the loop (despite choosing the word *circumference*, the loop need not be circular). Ampère's law then reads $BC = \mu_0 I_{enc}$, where I represents electric current and $\mu_0 = 4\pi \times 10^{-7}$ N/A^2 is a proportionality constant, called the permeability of free space.

The magnetic field lines for a long, straight, wire carrying a steady current make sets of concentric circles around the wire. By symmetry, the magnetic field is constant in magnitude over one of these circles. Since the circumference of a circle of radius r is $2\pi r$, Ampère's law states that $B 2\pi r = \mu_0 I_{enc}$. Assume that the wire is shaped like a right-circular cylinder of radius R_0. Outside of the wire, where the Ampèrian loop has radius $r > R_0$, all of the electric current is enclosed, so $B = \dfrac{\mu_0 I}{2\pi r}$. Inside the wire, where $r < R_0$, the Ampèrian loop encloses a fraction of the electric current. The electric current enclosed is proportional to the cross-sectional area of the wire: $\dfrac{I_{enc}}{I} = \dfrac{A_{enc}}{A}$. Since the area of a circle of radius r is πr^2, $I_{enc} = \dfrac{r^2}{R_0^2} I$. Thus, inside the wire, $B = \dfrac{\mu_0 I r}{2\pi R_0^2}$.

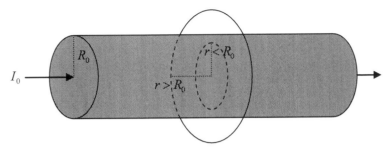

Ampèrian loops for a long, cylindrical wire carrying a steady current

According to Faraday's law, a changing magnetic flux can induce a current in a loop of wire. The direct way to supply current to a loop of wire – e.g. to power a lightbulb – is to connect a battery. Faraday's law provides an amazing alternative – a lightbulb can be powered by simply changing the magnetic flux in a coil of wire connected to the lightbulb, without any battery or direct power supply. Changing the magnetic flux requires one or more of the following: A time-varying external magnetic field, relative motion between a magnetic field and the loop(s) of wire, or changing the area of the loop(s) of wire. Simply waving a magnet near a loop of wire can induce a current in the wire – almost like a magic wand, except that the magic is just a fundamental concept of physics. Stokes's theorem is involved in inducing the current. The changing magnetic flux gives rise to electric field lines. When a loop of wire is in the presence of these electric field lines, the electric field accelerates charged particles in the conducting wire, giving rise to current. The line integral of the electric field over the loop of wire equals the induced emf, which is proportional to the induced current.

> **Gauss's law relates an integral of a field over a closed surface (the net flux) to an integral over an open volume bounded by the surface (the net source enclosed).**
>
> **Analogously, Stokes's theorem relates a line integral of a field over a closed curve to an integral over an open surface bounded by the curve.**
>
> **When applied to Ampère's law, Stokes's theorem states that the line integral of magnetic field over**

any Ampèrian loop is proportional to the net current enclosed by the Ampèrian loop.

When applied to Faraday's law, Stokes's theorem states that the line integral of electric field over a closed loop of wire (proportional to the induced current) is proportional to the instantaneous time rate of change of the magnetic flux. Thus, a changing magnetic flux can induce a current in a loop of wire.

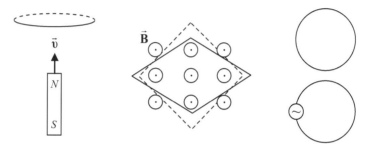

three examples of Faraday's law: a magnet approaches a loop of wire (left), inducing a current in the loop; a rhombus-shaped loop of wire compresses in the presence of a uniform external magnetic field (middle), inducing a current in the wire; and an alternating current in the bottom loop creates a time-varying magnetic field in the top loop (right), inducing a current in the top loop

7.5 Higher-Dimensional Gravity

The gravitational field is observed to satisfy both Newton's law of universal gravitation and Gauss's law in the known universe; the two laws express the same underlying principle in different forms. Gauss's law shows that the gravitational field depends largely on the geometry of the source. (Einstein went a step further, showing that gravity depends upon the geometry of spacetime, rather than just space, in his theory of general relativity.) One or more extra dimensions would have a profound geometric effect, and through Gauss's law could significantly affect the behavior of gravity.

Consider separately a point-mass (0D), an infinite line of mass (1D), and an infinite plane of mass (2D). The dimensionality of the source affects the flux of gravitational field lines: For the point-mass, the gravitational field lines radiate inward in three independent directions; for the infinite line, the gravitational field lines radiate inward in two independent directions (perpendicular to the line); and for the infinite plane, the gravitational field lines are perpendicular to the plane, heading into the plane. The net flux is greatest for the point-mass, where the field lines have the most freedom, giving an inverse-square law ($1/r^2$); the infinite line of mass yields a $1/r$ gravitational field; and the infinite plane results in a constant gravitational field.

Like the spherical planet, the gravitational field lines of a point-mass are radially inward. The gravitational flux through a Gaussian sphere concentric with the point-mass is $4\pi gr^2$ and the mass enclosed is simply that of the point mass, m. Gauss's law yields $4\pi gr^2 = 4\pi Gm$, which simplifies to $g = G\dfrac{m}{r^2}$, which is identical to the gravitational field outside of a spherical planet.

The gravitational field lines of an infinite rod are directed into the axis of the rod. A Gaussian sphere would not be practical here, but a cylinder nicely matches the symmetry of the field lines. The surface of a finite, right-circular cylinder includes two circular ends and a body. The flux through each end is zero because the gravitational field is parallel to the ends. The flux through the body is gA, where $A = 2\pi rL$ is the surface area of the body of the cylinder (which can be unfolded into a rectangular sheet with width $2\pi r$ equal to the circumference of the cylinder and length L). The mass enclosed is not the entire mass of the rod, but the mass m_L of that length of the rod inside of the Gaussian cylinder. Gauss's law gives $2\pi grL = 4\pi Gm_L$, which simplifies to $g = \dfrac{2Gm_L}{rL} = \dfrac{2G\lambda}{r}$, where λ is the mass per unit length of the rod. Cylindrical symmetry leads to a $1/r$ field, whereas spherical symmetry leads to a $1/r^2$ field.

The gravitational field lines for an infinite plane are perpendicular to the plane, directed toward the plane. A Gaussian cylinder turns out to be a good match, provided that the ends are parallel to the plane. In this case, the flux through the body is zero because the gravitational field lines are tangent to the body. The flux through each end is $gA = \pi gr^2$, so the net flux is $2\pi gr^2$. Gauss's law gives $2\pi gr^2 = 4\pi Gm_A$, where the mass enclosed m_A by the Gaussian cylinder is proportional to the area: $m_A = \sigma \pi r^2$, where σ is a constant equal to the mass per unit area of the plane. Thus, the gravitational field for an infinite plane is $g = 2\pi G\sigma$. Unlike cylindrical or spherical symmetry, the infinite plane has a constant gravitational field.

Let us compare the point-source, infinite line of charge, and infinite plane of charge. In the case of the point-source, the gravitational flux experiences the full freedom of 3D space, radiating inward three-dimensionally, and the gravitational field is proportional to $1/r^2$. For the infinite line of charge, the source itself spans an entire dimension, and the field lines radiate inward two-dimensionally; i.e. the source itself effectively reduces the flux from 3D to 2D. In this case, the gravitational field is proportional to $1/r$. The infinite plane of charge spans two dimensions, effectively leaving one dimension free, and the gravitational field is proportional to $1/r^0$, which is a constant because $r^0 = 1$.

Thus, simply by examining Gauss's law in ordinary 3D space, we are able to deduce that the power of r in the equation for gravitational field is very sensitive to the dimensionality of the flux of field lines. This way, we can see how the presence of one or more extra dimensions in our universe would significantly increase the flux of field lines in space, and significantly alter the power of r in the equation for gravitational field. When applied to a point-source, this corresponds to the power of r in Newton's law of universal gravitation or Coulomb's law.

In 3D space, Newton's law of universal gravitation is an inverse-square law, in agreement with Gauss's law. In higher dimensions, the power of r in Newton's law of universal gravitation must change in order to agree with Gauss's law. This agreement can be achieved if the force between two massive objects m_1 and m_2 with center-to-center separation r in N-dimensional space is $F = G_N \frac{m_1 m_2}{r^{N-1}}$, where both the numerical value and SI units of G_N depend upon the dimensionality of the space (the SI units of G_N are Nm^{N-1}/kg^2, where the italicized N represents the dimensionality while the normal N is a Newton). For example, in 4D space this would be an inverse-cube law. From a model-building perspective, in higher dimensions it is much more plausible to demand agreement between Newton's law of universal gravitation and Gauss's law, and therefore modify the power of r for gravity, than it would be to insist upon an inverse-square law. For one, demanding an inverse-square law would mean abandoning Gauss's law. Gauss's law is the mathematically elegant, conceptually powerful, and aesthetically desirable type of physical law that a model-builder would hope to be part of an ultimate theory of nature. Superstring theory – the chief motivation for considering extra dimensions and the best available candidate for a grand unifying theory of the universe – is consistent with Gauss's law, but not an inverse-square law, in higher dimensions. Gauss's law is also the basis for half of Maxwell's equations in electromagnetism.

In N-dimensional space, the net flux of field lines is found through an $(N-1)$-dimensional Gaussian hypersurface. For example, for a point-mass in 4D space, a useful Gaussian hypersurface would be the hypersurface of a glome because the gravitational field lines would radiate inward through 4D space, passing through the glome perpendicularly, and by hyperspherical symmetry the gravitational field would be constant throughout the hypersurface of the glome.

The charge enclosed is computed by integrating over the N-dimensional hypervolume enclosed by the $(N-1)$-dimensional hypersurface. Thus, Gauss's law expresses a relation between the N-dimensional hypervolume enclosed and the surface area of the bounding $(N-1)$-dimensional hypersurface.

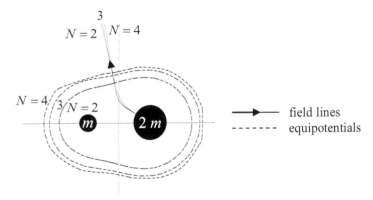

the effect that the dimensionality of space has on the shape of a field line and an equipotential for a binary system: the field line bends more sharply in higher dimensions; of course, the field lines also travel through 3D and 4D space for $N = 3$ or 4, whereas they only lie in the plane for $N = 2$

7.6 Higher-Dimensional Electromagnetism

Maxwell's four equations unify the seemingly differently behaving electric and magnetic fields into a single electromagnetic field. The seemingly separate effects of electricity and magnetism are inherently related, and are two different manifestations of the electromagnetic field. In integral form, two of Maxwell's equations involve Gauss's law and two involve Stokes's theorem:

$$\oint_S \vec{E} \cdot d\vec{A} = 4\pi k q_{enc} \qquad \oint_C \vec{E} \cdot d\vec{s} = -\frac{\partial}{\partial t} \int_S \vec{B} \cdot d\vec{A}$$

$$\oint_S \vec{B} \cdot d\vec{A} = 0 \qquad \oint_C \vec{B} \cdot d\vec{s} = \mu_0 I_{enc} + 4\pi k \mu_0 \frac{\partial}{\partial t} \int_S \vec{E} \cdot d\vec{A}$$

The top left equation is Gauss's law in electricity. The bottom left equation is Gauss's law in magnetism, where the right-hand side is zero because magnetic monopoles have never been observed. The top right equation is Faraday's law: The left-hand side represents the induced emf, which is proportional to the induced current, while the right-hand side is the instantaneous change in the magnetic flux. That is, a changing magnetic flux can induce a current in a loop of wire. The last equation is Ampère's law. The extra term in Ampère's law states that a changing electric flux can create a magnetic field, just like Faraday's law shows that a changing magnetic flux can create an electric field. Faraday's law would include a term analogous to $\mu_0 I_{enc}$ if magnetic monopoles existed. This makes the notion of a magnetic monopole very tempting theoretically as it would complete the symmetry in Maxwell's equations.

The net flux, e.g. $\oint_S \vec{E} \cdot d\vec{A}$, over a closed surface is distinctly different from $\int_S \vec{E} \cdot d\vec{A}$, which is the flux through an open surface. The flux through a sheet of paper, for example, is the flux through an open surface. The net flux through a sphere is the flux through a closed surface (in 3D space – in 4D space, a sphere is an open surface, while the hypersurface of a glome is closed). However, the flux through part of a sphere in 3D, such as a hemisphere, is the flux through an open surface.

As with Newton's law of universal gravitation, Coulomb's law must be transformed in higher dimensions in order to agree with Gauss's law, and hence to agree with Maxwell's equations that describe electromagnetism. In particular, Coulomb's law must be $F = k_N \frac{q_1 q_2}{r^{N-1}}$. Gauss's law has the same form in electricity as it does in gravity, relating an $(N-1)$-dimensional closed Gaussian hypersurface to the source enclosed.

Stokes's theorem analogously changes form in higher dimensions. In N-dimensional space, the usual line integral of the field becomes an integral over an $(N-2)$-dimensional closed Ampèrian hypersurface. For example, for a long, straight, current-carrying conductor in 4D space, the usual Ampèrian loop is replaced by a surface. If the current runs along the z-axis, a useful choice would be a sphere in the wxy hyperplane concentric with a point on the z-axis because every point on the sphere is equidistant from any given point on the z-axis. The current enclosed is computed by integrating over the $(N-1)$-dimensional hypervolume enclosed by the $(N-2)$-dimensional hypersurface. Thus, Stokes's theorem expresses a relation between the $(N-1)$-dimensional hypervolume enclosed and the hypersurface area of the bounding $(N-2)$-dimensional hypersurface. Compare to Gauss's law, which relates the N-dimensional hypervolume enclosed and the hypersurface area of the bounding $(N-1)$-dimensional hypersurface.

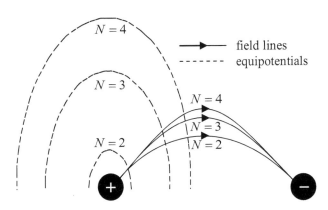

the effect that the dimensionality of space has on the shape of a field line and an equipotential for the electric dipole: the field line bends more sharply in higher dimensions; of course, the field lines also travel through 3D and 4D space for $N = 3$ or 4, whereas they only lie in the plane for $N = 2$

Geometrically, Gauss's law and Stokes's theorem naturally extend to higher dimensions. Since half of Maxwell's equations involve Gauss's law and half involve Stokes's theorem, Maxwell's equations also naturally extend to higher dimensions. However, there is a caveat, which requires mathematically expressing these equations in a somewhat different form: The magnetic field must be regarded as an anti-symmetric, second-rank tensor in higher dimensions instead of a pseudovector. This is related to the fact that cross products are only physically meaningful in 3D space. The usual cross product formulas can alternatively be expressed with one matrix and two vectors, and the matrix formulation does generalize to higher dimensions.

The electric and magnetic fields form the electromagnetic stress tensor. Maxwell's equations can be written concisely in terms of the electromagnetic stress tensor. This formulation shows how the electric and magnetic fields can be regarded as components of a single electromagnetic field. The temporal components of the electromagnetic field form the electric field, and the spatial components form the magnetic field. In $(3+1)$-D spacetime, each field has 3 components:

$$\begin{pmatrix} 0 & E_x/c & E_y/c & E_z/c \\ -E_x/c & 0 & B_z & -B_y \\ -E_y/c & -B_z & 0 & B_x \\ -E_z/c & B_y & -B_x & 0 \end{pmatrix}$$

In $(N+1)$-D spacetime, the electric field has N components – the same number as a vector – and the magnetic field has $\dfrac{N(N-1)}{2}$ components – the same number as an anti-symmetric, second-rank tensor. For example, in $(4+1)$-D spacetime,

$$\begin{pmatrix} 0 & E_1/c & E_2/c & E_3/c & E_4/c \\ -E_1/c & 0 & B_{12} & B_{13} & B_{14} \\ -E_2/c & -B_{12} & 0 & B_{23} & B_{24} \\ -E_3/c & -B_{13} & -B_{23} & 0 & B_{34} \\ -E_4/c & -B_{14} & -B_{24} & -B_{34} & 0 \end{pmatrix}$$

Gauss's law, associated with the dot product between vectors (the field and the differential area element), naturally extends to higher dimensions both algebraically and geometrically. Stokes's theorem, the analogous law associated with the cross product between vectors (the field and the differential arc length), naturally extends to higher dimensions geometrically, as we have seen. Algebraically, Stokes's theorem also extends to higher dimensions naturally in its most elegant form, expressed as an integral in differential geometry. This relates to the point made in Chapter 6 – that physical laws traditionally expressed as cross products in 3D can be generalized to higher dimensions either through the advanced technique of differential geometry or the more straightforward approach of expressing the pseudovector (in this case, magnetic field) as an anti-symmetric, second-rank tensor and rewriting the cross product as a matrix multiplication, even though the cross product itself is only physically meaningful in 3D.

All of the laws of physics that can thus be viably generalized to higher dimensions. Those formulas that normally involve a cross product can either be recast in the form of differential geometry or matrix multiplication with an anti-symmetric, second rank tensor. Even very geometric laws, like Gauss's law and Stokes's theorem, naturally extend to higher dimensions, but in this case they have an inherently different form, with an increase in the power of r in the denominator.

7.7 Nuclear Forces

Atoms consist of a positively charged nucleus, where protons and neutrons reside, surrounded by electron clouds. A typical atom has a size of about an Angstrom (1 Å = 10^{-10} m), while the size of a typical nucleus is on the order of 10^{-15} m. A typical atom is therefore about 100,000 times larger than its nucleus. Since the electrons are evidently pointlike particles, matter consists primarily of empty space.

The electrons surrounding the nucleus are attracted to the protons via the electromagnetic interaction, since opposite electric charges attract. This explains how the electrons are bound to the atom. However, think about the protons and neutrons residing in the tiny nucleus. There may be up to a hundred or sore protons, and even more neutrons, occupying this very tiny space. The protons repel one another electrically, so what possesses these particles that loathe one another electrically to stay in the nucleus? It turns out that protons and neutrons attract each other via the strong nuclear force. At very short range, such as the size of an atomic nucleus, the attraction of the protons to neutrons and other protons via the strong nuclear force overpowers the electrical repulsion of the protons. The strong nuclear force is responsible for binding protons and neutrons together in the nucleus.

Puzzle 7.15: The lighter elements have nearly the same number of neutrons as protons, but the heavier elements have many more neutrons than protons. Why?

Another nuclear force – the weak nuclear force – is responsible for radioactive decays. Some elements have isotopes with unstable nuclei which can decay to stable nuclei, usually by emitting alpha particles (helium nuclei), beta particles (electrons), or gamma rays (photons). The weak nuclear force mediates such decays, including the radioactive decay of the uranium isotope $^{238}_{92}U$, which emits a relatively large amount of energy in nuclear fission, and the carbon isotope $^{14}_{6}C$, for which the half-life of 5700 years is useful in radiometric dating.

The four fundamental forces of nature include the gravitational force, the electromagnetic force, the strong nuclear force, and the weak nuclear force. The gravitational and electromagnetic interactions are observed to be inverse-square laws with apparently infinite range. Any two massive objects in the universe exert a gravitational force on one another – even the most distant galaxies – although the magnitude of the gravitational force decreases rapidly as $1/r^2$. Coulomb's law has similar form, where apparently the r in $1/r^2$ has infinite range, though it is much more difficult to test that this holds over astronomical distances. The strong and weak nuclear forces, however, are very short-range forces – they are strong in their domain, which is about the size of an atomic nucleus, but do not have far-reaching effects like the gravitational and electromagnetic forces.

The strong nuclear force is about 100 times stronger than the electromagnetic force for interaction distances up to about 10^{-15} m. The strong nuclear force is about 10 trillion times stronger than the weak nuclear force. Beyond the nuclear regime, the electromagnetic force is the dominant interaction. The electromagnetic force between two charged particles is typically incredibly strong compared to the gravitational attraction between them. For example, the mutual repulsion between two electrons is about 10^{43} times stronger than their gravitational attraction.

Puzzle 7.16: Which of the fundamental forces is primarily involved in the clapping of two hands together?

Volume 2: The Physics of the Fourth and Higher Dimensions...

Further Reading:

The 3D physics of this chapter, including gravitational fields, Gauss's law, Stokes's theorem, electricity and magnetism, and the four fundamental forces of nature, can be explored in much more detail in an introductory physics textbook. An advanced reader with a very strong mathematical repertoire can best understand how to geometrically extend Gauss's law to higher dimensions by studying differential geometry, which is a more sophisticated way to generalize the cross product than was considered in Chapter 6. The intermediate reader who has a background in tensors, but who does not want to jump into differential geometry, may be interested in an unpublished work [T1] that applies the tensor constructions described in Chapter 6 to generalize Maxwell's equations without applying the techniques of differential geometry.

8 Compactification

Now we will explore how ordinary particles can propagate into one or more extra dimensions without contradicting our experience, which suggests that our universe is 3D. In particular, the extra dimensions could be hidden if they are compact. We will address the issues of *how* and *why* one or more dimensions may be compact. We will also explore the size and shape of extra dimensions. Regarding the size, this includes the original prediction by superstring theory and recent motivation for superstring-inspired extra dimensions to be much larger than originally thought. Finally, we will take a brief survey of higher-dimensional models, which we will explore from an experimental perspective in the following chapter.

8.0 Non-Euclidean Geometry

In a Euclidean space, the Cartesian coordinates x, y, z, etc. can assume any real value (e.g. $-\infty < x < \infty$) independent of the other coordinates. A Euclidean space is an infinite line in 1D, an infinite plane in 2D, an infinite 3D hyperplane in 3D, an infinite 4D hyperplane in 4D, etc. In a Euclidean space, the shortest distance between two points is a straight line.

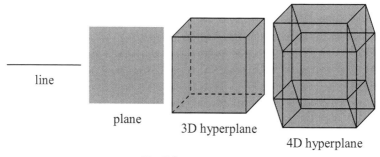

Euclidean spaces

Curved spaces are non-Euclidean. For example, the surface of a sphere is a non-Euclidean space. A universe shaped like a sphere, for which objects can only move along the surface of the sphere, would be 2D. Objects moving in such a universe have only two degrees of freedom. In fact, 2D beings inhabiting such a universe might believe that their universe is planar, rather than spherical, especially if the radius is very large. On the other hand, if the radius were small enough, one interesting observable feature would be that an object could travel in one direction and return to its starting point without ever changing direction.

> In this chapter and onward, we refer to the dimensionality of a manifold by the number of independent coordinates needed to describe it. For example, a sphere is a 2D surface and a glome is a 3D hypersurface. (Note that this is different from the convention we used in Chapter 4.[4])

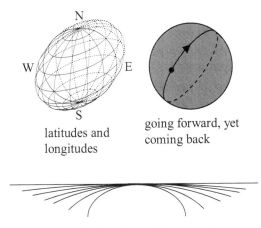

latitudes and longitudes

going forward, yet coming back

a larger sphere looks flatter when viewed up close

The space of our universe is a manifold, which may be non-Euclidean and may consist of more than three dimensions. If it is a 3D Euclidean space, then the manifold is a 3D hyperplane. However, the manifold may be a cubinder, for example, with three noncompact dimensions and one compact dimension. In this case, the manifold is a

---

[4] The reason for this difference is that in the context of compactification we are concerned with the manifold, where the dimensionality of the manifold is determined by the number of independent directions into which a particle could propagate if confined to the manifold. In Chapter 4, the context was different: There it was deemed more intuitive for a reader to adopt a Euclidean mindset.

4D space, meaning that a particle confined to the manifold can travel in any linear combination of four independent directions. It would require a 5D Euclidean graph to plot such a cubinder, but since the interior of the circular dimension is not accessible to any particles, only four of the Cartesian coordinates are independent.

For Euclidean manifolds, such as the line and the plane, the dimensionality of the manifold equals the number of Cartesian coordinates needed to describe the manifold. For non-Euclidean manifolds, not all of the Cartesian coordinates needed to map out the manifold are independent, and so the dimensionality of the manifold is less than the number of Cartesian coordinates needed to describe it. There do exist other coordinate systems where the number of independent coordinates equals the dimensionality of the space. For example, three Cartesian coordinates are needed to unambiguously specify the location of every point on the surface of a sphere, but these coordinates are not independent since $z = \sqrt{R^2 - x^2 - y^2}$. In spherical coordinates, only two coordinates (longitude and latitude) are needed to fully specify the location of every point on the surface of a sphere. (In both cases, the radius is not a variable, since it is a constant for any given sphere.)

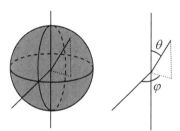

the two independent coordinates
of a spherical manifold

The sphere is a 2D manifold with two compact dimensions. The shortest distance between any two points on the sphere is the great circle that connects them. If you are used to thinking in Euclidean terms, you may want to say that if you connect the two points with a straight line the distance will be shorter than if you connect them with a circle. However, it is not possible to take this straight-line route without leaving the space of this curved 2D manifold – i.e. the surface of the sphere.

Volume 2: The Physics of the Fourth and Higher Dimensions...

the shortest between any two points in a spherical manifold is a great circle; it is not possible for a particle confined to the manifold to travel along a straight line connecting the two points

The right-circular cylinder is a 2D manifold with one compact dimension and one Euclidean dimension. A particle confined to this 2D space can travel off to infinity along the noncompact dimension, while motion along the compact dimension is periodic. The shortest distance between two points on this manifold is a straight line, a circle, or a helix. If the axis of the cylinder is the z-axis, then the shortest distance between two distinct points is a circle if both points share the same value of z, a straight line if they share the same values of both x and y, and a helix otherwise.

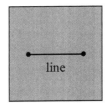 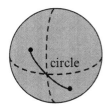 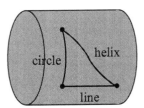

the shortest distance between two points for various 2D manifolds

The geometry governing a non-Euclidean space can be vastly different from the geometry of Euclidean space. For example, in Euclidean space, the sum of the interior angles of a triangle equals 180°. For a sphere, the equivalent "tri-angle" consists of three points connected by arcs of great circles, and the sum of the interior angles equals 270°.

Here is another example. In a 2D plane, if you walk one mile south, one mile east, and one mile north, you will always wind up one mile east of where you started. However, if you try this on the surface of a sphere, the answer is generally different. For example, if you start from the North Pole and try this, you will wind up exactly where you started. In this case, it is easy to see that an equilateral "tri-angle" on the surface of a sphere has interior angles of 90° (since north and east are perpendicular), which agrees with the rule that the three angles must sum to 270°.

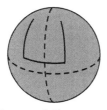
traveling equal distances south, east, and north on a sphere

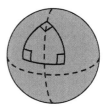
an equilateral "tri-angle" on a sphere has right angles

Puzzle 8.1: On the surface of a sphere, there is an infinite set of points from which you can travel one mile south, one mile east, and one mile north and wind up where you started. One answer is the North Pole. Where are the other possible starting points?

8.1 Spacetime Curvature

If the notion that one or more dimensions of our universe may be curved seems strange to you – as would be the case if the space of the universe is a non-Euclidean manifold – then consider this: It is well-known that spacetime is curved in our universe. The three dimensions of space and one dimension of time that are readily observable in our universe are woven together to form spacetime. Upon very close inspection, the spacetime of our universe actually curves.

This follows from Einstein's general theory of relativity, which is described in more detail in Chapter 10. However, we will consider a few ideas here that will be helpful for understanding the curvature of spacetime.

Light, which is an electromagnetic wave, consists of packets of energy. Each packet of energy is a particle of light, called a photon. A photon has zero rest-mass, but photons are never at rest: Photons travel the speed of light, c, which equals 2.9979×10^8 m/s in vacuum. Since a photon is a packet of electromagnetic energy, it has relativistic mass according to Einstein's famous equation, $E = mc^2$.

The rest-mass of a particle, such as an electron, is how much inertia it has relative to a reference frame where the particle is at rest. In a reference frame where the particle is moving, it has more inertia –

called *relativistic mass*. Einstein's equation expresses an equivalence between the particle's energy and relativistic mass: $E = mc^2$. In terms of the particle's rest-mass m_0, its energy equals $E = \sqrt{p^c c^2 + m_0^2 c^4}$, where $\vec{p} = m\vec{v}$ is the particle's relativistic momentum (relativistic mass times velocity). The relativistic mass and rest-mass are also related through the equation $m = m_0 / \sqrt{1 - v^2/c^2}$. In the limit that v approaches zero, the relativistic mass equals the rest-mass and the energy equals $m_0 c^2$.

A photon travels the speed of light (so $v = c$) and has no rest-mass (i.e. $m_0 = 0$), but does have relativistic mass and carries energy and momentum: $E = mc^2$ and $E = pc$. Recall (from Chapter 7) that mass serves not only as a measure of inertia, but also as the source of a gravitational field. Photons interact with gravity through their relativistic mass. Gravity actually bends light since any mass, including the relativistic mass of a photon, experiences a force in the presence of a gravitational field.

Einstein's theory of general relativity predicts the curvature of a beam of light passing through a gravitational field. This has been confirmed experimentally, but the effect is generally quite subtle: You won't notice the effect if you try to detect the free fall of a laser beam in the same way as you would for a projectile. This is illustrated by comparing the following examples.

Consider a rifle that is mounted horizontally 100 m from a wall. A bull's-eye is placed on the wall at the same height as the rifle. When the trigger is pulled, the bullet is launched with an initial velocity of 500 m/s. Neglecting air resistance, where does the bullet strike the wall?

Once the bullet leaves the barrel of the rifle, the only force acting on it is the earth's downward gravitational pull. The rifle only exerts a force on the bullet while the bullet accelerates through the barrel. Afterward, the bullet experiences no horizontal forces, but it does have inertia – a natural tendency to continue moving horizontally. Thus, the bullet maintains its horizontal component of velocity of 500 m/s. The bullet also develops a vertical component of velocity due to gravity's downward pull – vertically, the bullet gains 9.8 m/s of speed each second.

Since the bullet continues to travel 500 m/s horizontally and the wall is 100 m away, the bullet strikes the wall in $t = x/v_x = 100 \text{ m}/500 \text{ m/s} = 0.2 \text{ s}$. During this time, the vertical component of the bullet's velocity increases. The bullet is initially heading horizontally, so the initial vertical component of its velocity is zero: $v_{y0} = 0$. Since it gains 9.8 m/s of speed vertically each second, when the bullet strikes the wall the vertical component of its velocity is $v_y = -gt = -9.8 \text{ m/s}^2 \cdot 0.2 \text{ s} = -1.96 \text{ m/s}$. The minus sign signifies that gravity is pulling downward on the bullet. On average, the vertical component of the bullet's velocity is $\bar{v}_y = (v_{y0} + v_y)/2 = (0 - 1.96 \text{ m/s})/2 = -0.98 \text{ m/s}$. Thus, the bullet descends a vertical distance of $y = \bar{v}_y t = -0.98 \text{ m/s} \cdot 0.2 \text{ s} = -0.196 \text{ m}$. That is, the bullet misses the target by 19.6 cm due to gravity's downward pull.

gravitational deflection of a projectile

Now let us replace the rifle with a laser. So now a laser is mounted horizontally 100 m from a wall. When the laser is turned on, where does the light strike the wall?

The difference between light emitted by the laser in this example compared to the bullet shot by the rifle in the previous example is that light travels 3.00×10^8 m/s through air, whereas the bullet was launched with a speed of 500 m/s. In one second, light would travel three hundred million (300,000,000) meters, but the bullet would travel a mere five hundred meters this same time. In both cases, the path is curved, but while the bullet deviates significantly from its initial heading (landing 19.6 cm below its target), the effect is minute for the laser light.

The reason is that the bullet spends 0.2 s in the air, so gravity acts on the bullet long enough for it to gain an appreciable downward component of velocity (0.98 m/s). On the other hand, the laser light travels the horizontal distance of 100 m in a few microseconds: $t = 100 \text{ m}/3.00 \times 10^8 \text{ m/s} = 3.33 \times 10^{-7} \text{ s}$. During this time, light gains a downward component of velocity of just $v_y = -9.8 \text{ m/s}^2 \cdot 3.33 \times 10^{-7}$ s $= -3.27 \times 10^{-6}$ m/s. The average vertical component of the velocity of the laser light is $\bar{v}_y = (-3.27 \times 10^{-6} \text{ m/s})/2 = -1.64 \times 10^{-6}$ m/s. Thus, over the course of the 100 m distance traveled horizontally, light descends a mere $y = -1.64 \times 10^{-6}$ m/s $\cdot 3.33 \times 10^{-7}$ s $= -5.44 \times 10^{-13}$ m. That's over a hundred times smaller than the size of an atom! Here, we have assumed naïvely that we can apply the same mathematics to the laser light, instead of looking to Einstein's general theory of relativity, yet this rough calculation effectively illustrates the main concept.

Although gravity does bend light as a result of light's relativistic mass, the effect is terrestrially insignificant compared to gravity's effect on bodies that do have rest-mass. It takes an object with a much stronger gravitational field to create a noticeable effect. The sun, for example, behaves as a gravitational lens for starlight passing nearby. Einstein predicted that starlight passing close to the sun's surface should deflect about two arcseconds as viewed from earth. Although this is very slight – one arcsecond equates to one 3600th of a degree – such shifts have indeed been measured. A distant star traveling with constant speed develops a non-uniform acceleration relative to observatories as it passes behind the sun due to the sun's warping of spacetime. This is strong experimental evidence in favor of Einstein's general theory of relativity.

It is not just that light curves, but that spacetime itself is warped. Any object, including light, traveling through spacetime experiences the effect of spacetime curvature. Ignoring the effects of gravity, light would travel in straight-line paths (well, we'll save the uncertainty principle for Chapter 10, where we introduce concepts from quantum mechanics). If you draw a line on a sheet of paper, it looks straight – unless you bend the sheet of paper, in which case the line curves. This is how we think of light: Light travels in a straight line on the manifold of spacetime, but very massive objects cause spacetime itself, and hence light, too, to curve.

It's not just an appealing conceptual tool – this is how the mathematical theory is worked out. For three dimensions of space and one of time, we begin with the Cartesian variables x, y, z, and t – as if spacetime were flat, but then spacetime becomes curved when the source of a gravitational field is introduced. Light travels along straight-line paths[5] before the source of a gravitational field is added, and bends after the manifold of spacetime is warped by the presence of such a source.

The effect is much more drastic in the case of a black hole. Inside the event horizon, the gravitational pull of the black hole is so strong that even light itself cannot escape. Indirect observation of

---

[5] In Euclidean space; light also bends when traveling along a curved manifold that results from compactification, as well as from gravity.

black holes serves as experimental confirmation that spacetime can be very noticeably warped.

Since a black hole can warp spacetime so much that it can trap light – and hence anything – inside of the event horizon, it is not such a huge leap to fathom spacetime being warped so much that one or more dimensions could be compactified. Consider all of the energy of the universe confined to a small space just after the Big Bang, and recall that energy is equivalent to relativistic mass through $E = mc^2$. If a black hole can create an event horizon, is it unreasonable for one or more dimensions to warp so much that they become compact?

Perhaps this is not the only, or even the best, means for compactification, but this conceptual argument does show that the idea of compactification is not a giant leap from our knowledge of spacetime curvature. We will return to the curvature of spacetime and black holes in Chapter 10.

8.2 Hidden Dimensions

The known universe appears to have just three spatial dimensions, yet superstring theory predicts the existence of six additional dimensions. For superstring theory to be viable, the obvious question is: Where are the extra dimensions? Evidently, they cannot be Euclidean dimensions into which protons, neutrons, and electrons can propagate – otherwise, macroscopic objects could move in any combination of four independent directions. Nor is it possible for gravitons, which mediate the gravitational interaction, or photons, which mediate the electromagnetic interaction, to have full freedom in an extra Euclidean dimension because the increased flux would change the power of r in Gauss's law for gravity, electricity, and magnetism.

Thus, if there are extra dimensions, they must be hidden in some way. They must at least be hidden in such a way that Gauss's law would behave very nearly as an inverse-square law from a macroscopic perspective; and if the particles that make up ordinary matter can travel in the extra dimensions, then the extra dimensions must also be hidden such that they would not be readily detected (which is not to say undetectable) by macroscopic beings – in particular, the nature of any hidden dimensions must explain why it does not seem possible to intelligent, macroscopic beings to travel into an extra dimension.

It is possible that there are extra dimensions which are hidden through compactification. There may be three Euclidean dimensions and one or more compact dimensions. In the case of superstring theory, we expect six extra dimensions. In M theory, this is effectively increased to seven. We have superstring-inspired extra dimensions in mind since superstring theory is presently the most appealing candidate for a higher-dimensional theory, but we can work on more general grounds by allowing for any number of extra compact dimensions.

If the size of the extra compact dimensions is sufficiently small, this would explain why only three dimensions are readily observed by macroscopic beings. To see this, consider the simple case of a 2D universe with one Euclidean dimension and one compact dimension. This could come about as follows. The 2D universe could begin as an infinitesimal square, which grows into a larger square. At some point early in its development, one dimension may begin to curve. It could curve so much that the square bends into a right-circular cylinder. Put another way, one dimension remains noncompact, while the other is compactified. From this point onward, the noncompact dimension may grow – i.e. matter continues to spread out along the length of the cylinder – while the compact dimension may remain fixed in size. It is easy to make variations of this simple model, but it illustrates the basic notion of compactification.

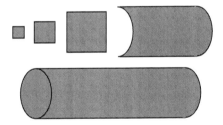

a simple 2D model for compactification: a 2D universe begins as an expanding plane; at some point, one of the dimensions compactifies; afterward, the noncompact dimension grows, while the compact dimension does not

Why might the compact dimension be hidden? We will see that this depends upon the radius of the cylinder's circular cross section. Imagine a macroscopic 2D being living on the surface of this cylinder. Such a being may consist of billions of 2D cells, which are themselves composed of billions of 2D atoms. Suppose that the 2D being is about a meter long in the Euclidean dimension. The 2D being could travel infinitely far along the Euclidean dimension, but is limited in motion along the compact dimension. Just how limited depends upon the radius of the compact dimension.

If the radius of the compact dimension is quite large – say, billions of light years – the space would appear flat – i.e. the 2D being would not realize that the space is curved. The earth similarly appears to be flat to humans, who are very small compared to the radius of the earth, although the spheroidal shape of the earth does have some observable effects. If the radius of the compact dimension is not quite as large – maybe just a few light years – the space would still appear locally flat, but one interesting consequence of the compactification would be that the 2D being could see its own star by looking in two different directions.

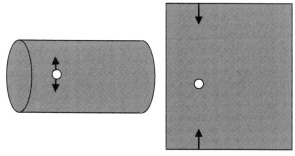

an object on the manifold of a 2D cylinder could see its own image, if light making the journey around the compact dimension could be seen (also shown for an unfolded cylinder)

Making the radius of the cylinder smaller – say on the order of a kilometer – the 2D being would be able to walk in a circle around the compact dimension, moving only in one direction yet able to return to its starting point. At this size, the curvature of the 2D being is quite evident, even though the 2D being would not see curves directly – not as a 3D being could see an obvious curve by looking at this manifold from a distance as illustrated here. To the 2D being, motion is very much like motion along a plane; the difference is that motion along one dimension is periodic.

It would be like an ant crawling along the surface of a cylinder. If the ant crawls along the surface of a very wide cylindrical tower, the ant experiences a rather planar motion; the ant may never realize that the tower is curved. If the cylinder is as wide as a soda can, the ant can travel in a complete circle, but otherwise experiences two degrees of freedom similar to crawling on a rectangle. If the ant crawls along a straw, the ant's body is almost equal to the circumference of the compact dimension. In this case, motion is nearly 1D. As we consider a thinner and thinner cylinder, in order to make the proper analogy with compactification we must not have a regular 3D ant, but must consider a 2D being lying exactly on the surface of the cylinder.

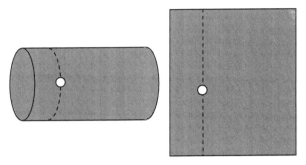

from the manifold's perspective, light appears to travel in a straight line – this is seen here in the unfolded cylinder; but when the cylinder is viewed externally from a Euclidean perspective, light traveling around the cylinder appears to follow a curved path

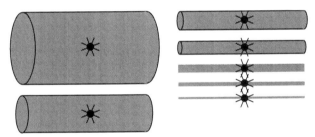

as the radius of compactification gets smaller and smaller compared to the size of a 2D being on its manifold, motion along the manifold becomes effectively more and more 1D from the 2D being's perspective

Returning to our hypothetical 2D being, there is an interesting possibility if the radius of the cylinder equals the width of the 2D being: the 2D being could be ringlike in shape. In this case, the 2D being experiences primarily 1D motion along the length of the cylinder, and a twisting motion achieved by rotating its ringlike structure. As we consider a smaller and smaller radius, the 2D being experiences more 1D motion and notices the presence of a second dimension less.

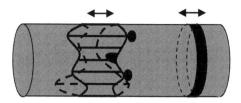

a 2D being that spans the compact dimension would experience 1D motion along the noncompact dimension in the 2D space – like a bead sliding along a wire

In the extreme limit that the extra dimension becomes smaller than the size of a 2D atom, the being itself begins to appear very 1D. If the being cannot see objects smaller than a tenth of a millimeter or so, how could it possibly notice a compact dimension smaller than a nanometer? It would not be able to directly notice elementary particles that make up its atoms moving through this compact dimension.

That is the spirit with which extra dimensions may be hidden in our universe. We will consider compactification in more detail here, and then consider how we might detect the presence of such extra dimensions in our universe in the next chapter.

8.3 Compactification Schemes

Three basic properties of compactification to consider are the number, size, and shape of the compact dimensions. These extra compact dimensions in conjunction with the three usual dimensions form the complete manifold.

The simplest case of compactification that can coincide with the observed universe corresponds to a single extra compact dimension. While superstring theory suggests six extra dimensions, and M theory effectively seven, much research is conducted with a single extra dimension in mind because this is much easier to visualize, the mathematics is much simpler, and this illustrates the key features of compactification. In Sec. 8.7, we will also see that there are plausible models in which one of the extra dimensions is significantly different from the others, in which case the other extra dimensions can often be neglected. However, depending upon the model and calculation, it can be crucial to consider all of the extra dimensions – not just one. We will return to this point later.

For the case of three Euclidean dimensions and a single compact dimension, the manifold of the universe would be the 4D hypersurface of a cubinder. The volume of such a cubinder would be 5D. Five Cartesian coordinates would be required to describe the 4D hypersurface of this cubinder: x, y, z, w, and υ. However, only four of these coordinates are independent, since w and υ form the compact dimension: $w^2 + \upsilon^2$ equals a constant (the radius of the circle squared). This manifold is 4D because it can be described with four independent coordinates (4D cylindrical coordinates): x, y, z, and θ, where θ is an angular variable that indicates the position of a point along the circular dimension. There are three Euclidean dimensions and one compact dimension.

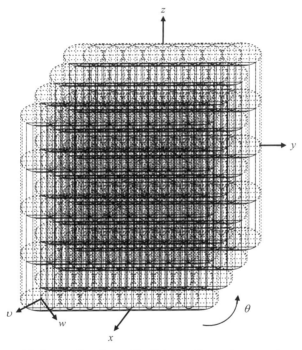

a cubinder with three Euclidean dimensions and one compact dimension is a 4D manifold; the single coordinate θ is equivalent to the two dependent variables υ and w

A particle on the 4D hypersurface of such a cubinder could move infinitely far in a straight line along x, y, or z – or any linear combination of these – which is akin to human experience, but it can also move along the compact dimension. A particle for which θ

changes while the x-, y-, and z- remain unchanged would appear stationary in the Euclidean dimensions, but would travel in a circle in the compact dimension. A photon traveling in such a circle would appear to be at rest to a macroscopic being, and would therefore behave like a massive stationary object (the rotational energy being equivalent to mass through Einstein's equation) and not like a photon at all. However, you would never see such a photon – it's not moving along the Euclidean dimensions, so it won't ever get to your eye. Well, suppose your eye runs into a such a photon – you still wouldn't notice anything unusual unless you happen to run into a very large number of such photons.

A particle could also travel along a helical path – moving in a combination of the Euclidean dimensions and the compact dimension. The particle would appear to travel in a straight line to a macroscopic being. A photon traveling along such a helix would be effectively slower than the speed of light and have more relativistic mass (again, from the rotational energy in the compact dimension) relative to a macroscopic being that only observes the Euclidean dimensions directly. We will return to this point in the following section.

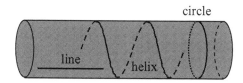

a photon moving along the noncompact dimension appears normal, while a photon moving in a circle along the compact dimension appears to be at rest and a photon moving in a combination of the two directions appears to be a heavier, slower-moving particle from a macroscopic, effectively 1D perspective

There is another interesting possibility to consider: Perhaps *all* of the dimensions are compactified! When you first learn about compactification, one of the first things you might wonder is why three dimensions are noncompact while others are compact. We addressed this issue when we considered spacetime curvature. However, we ignored the possibility that all of the dimensions, including the three known dimensions, are compact.

The three known dimensions could be compact, yet seem to be Euclidean from our perspective, if the radius of compactification of the known dimensions is extremely large on the astronomical scale. The three known dimensions may be three independent dimensions of the

3D hypersurface of a hypersphere, for example. Such a hypersphere would encompass a 4D volume, but the manifold would be 3D because the inward/outward direction would be inaccessible to an object confined to the hypersurface.

It seems that radius of the 3D hypersphere would need to be very large compared to the size of the Milky Way galaxy. For one, this is necessary to explain why we do not see mirror images of our Milky Way galaxy at the horizon of the universe. (You would actually see light from our galaxy in *any* direction you look, since photons from our galaxy would make a roundtrip, returning to our galaxy regardless of which way they were emitted – except for those that are intercepted, or if the circumference is so large that the first photons haven't yet completed a roundtrip. However, the galaxy may have accelerated significantly since emitting the first batch, which is something else to consider.) There could be other effects of such curvature, including cosmological implications, requiring the radius to be much larger than the span of the visible universe. As we consider a larger and larger radius, eventually any effects of the curvature of the hypersphere would become so slight that we would not be able to discern that the three known dimensions are not Euclidean.

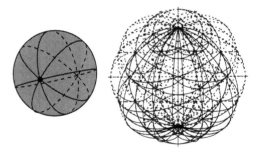

in a 2D universe on the surface of a small enough sphere, an object could potentially see itself by looking in any direction; the same would be possible in a 3D universe on the surface of a hypersphere

So, it is possible that the known dimensions are also compact. In this case, the known dimensions would have an extremely large size – much larger than the visible universe – while the compact dimensions are very tiny. The question becomes not why are some dimensions compact and others not, but why are some very large and others very tiny.

In the case where all of the dimensions are compactified, the complete manifold could be a hyperspheroid with three extremely large compact dimensions and one or more very tiny compact dimensions. We can visualize this as an extension of the surface of a 2D torus. A torus has two compact dimensions: One dimension along the circular axis of the torus, and one along the circular cross section.

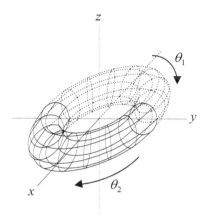

the two independent directions on the manifold of a torus

Of course, the notion that all of the dimensions are compact is just fun, interesting, and philosophical – at best – unless some theoretical motivation is developed or experimental observation discovers it. Well, there is one other reason, which we have utilized: It served as another example of compactification.

Let us return to the case where three dimensions are Euclidean, but now consider the impact of two or more extra dimensions. Multiple extra dimensions allow for a variety of manifolds of different shape. Three common compactification schemes with multiple extra dimensions include spherical compactification, toroidal compactification, and the Calabi Yau manifold.

Spherical compactification corresponds to a cubinder with three Euclidean dimensions and the remaining $\delta \geq 2$ dimensions compactified on the surface of an δ-dimensional sphere. The total number of dimensions of the complete manifold is $N = 3 + \delta$. A particle on this type of manifold can travel infinitely far in any linear combination of three Euclidean dimensions, and for any given coordinates (x, y, z) a particle can travel along the surface of a sphere specified by the two angles θ and φ. The radius of compactification is a fixed parameter, not a variable, and so does not contribute toward the dimensionality of the manifold. This is similar to the 4D cubinder considered previously for the case of one extra dimension except that a particle anywhere in the three known dimensions can travel along the surface of a sphere, rather than a circle, in the extra dimensions.

In spherical compactification, the extra dimensions are compactified on the surface of a δ-dimensional sphere.

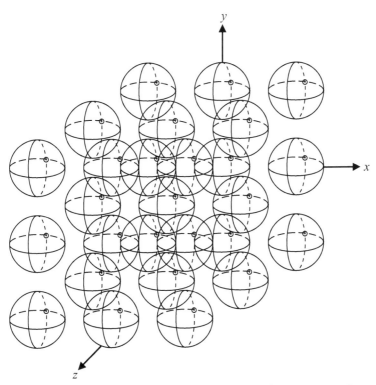

spherical compactification with two extra dimensions corresponds to a generalized spherinder; at any point (x,y,z) in 3D space, a particle can travel along the surface of a compactified sphere in addition to the usual three Euclidean directions

In the case of toroidal compactification, the $\delta \geq 1$ extra dimensions are compactified on the surface of a δ-dimensional torus.

The case $\delta = 1$ corresponds to the 4D cubinder with one compact dimension. For $\delta \geq 2$, a particle at any point in the usual 3D space can move along the known dimensions or along the surface of a δ-dimensional torus. Toroidal compactification with $\delta = 1$ is quite popular for its mathematical and visual simplicity, and illustrates key features of higher-dimensional models.

For toroidal compactification, the extra dimensions are compactified on the surface of a δ-dimensional torus.

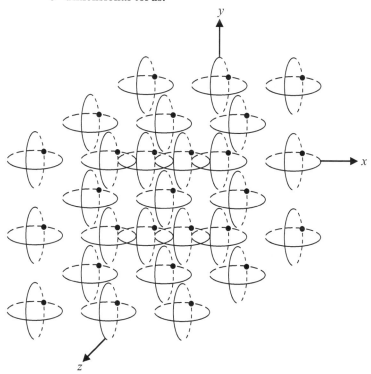

toroidal compactification with two extra dimensions corresponds to a generalized cubinder; at any point (x,y,z) in 3D space, a particle can travel along either of two mutually perpendicular compact circles in addition to the usual three Euclidean directions

The full six extra dimensions from superstring theory (or effectively seven from M theory) compactify on a Calabi Yau manifold; being 9D (or 10D), with three noncompact dimensions, this is a rather complicated structure, but still exhibits many of the key features highlighted by simple models with one or two extra dimensions.

8.4 Toroidal Compactification

Let us consider the case of a single, tiny extra dimension compactified on a torus such that the complete manifold is the 4D hypersurface of a cubinder. This simple model will illustrate some important features of compactification.

First, let us get acquainted with some notation. This will provide a shorthand means of describing directions in the 4D space, and will help to understand the graphs. In Cartesian coordinates, the manifold is described via the five variables x, y, z, w, and υ, where x, y, and z are Euclidean and w and υ form a circle according to the equation $w^2 + \upsilon^2 = R^2$, where R is the radius of the compact dimension. Note that R is a constant, not a variable. There are four independent variables, since w and υ are related through the constant R.

It is convenient to work with four independent variables in lieu of five Cartesian coordinates. We could work with the 4D hypercylindrical coordinates, x, y, z, and θ, where θ is an angular measure from the center of the circular dimension that specifies where a point is in the extra dimension. Alternatively, we could work with x, y, z, and s, where s is the arc length in the extra dimension. These two formulations are related since $s = R\theta$. The main difference is that s has the same units as x, y, and z, but θ does not.

We can gain some insight into the 4D cubinder by reviewing the right-circular cylinder, which is 2D. Let us consider a right-circular cylinder lying on the z-axis. This cylinder is fully specified by two independent coordinates, which we could call z and θ. Alternatively, we could choose z and s, where again $s = R\theta$. A third coordinate, though not independent, is needed in Cartesian coordinates: Here, we use x, y, and z, where $x^2 + y^2 = R^2$.

Observe that the cylinder can be unfolded into a plane after first cutting it along its length. In this way, the manifold of the cylinder can be mapped onto a plane. The cylindrical variables z and s essentially become Cartesian coordinates on the unfolded cylinder, except that s is periodic.

a particle traveling on the surface of a right-circular cylinder has two degrees of freedom

    Imagine drawing the following straight lines on this unfolded cylinder: one along z, one along s, and one with an in-between slope. If this plane is folded back into a cylinder, the line drawn along z will still be a line, but the line along s will turn into a circular arc and the other line will turn into an arc of a helix.

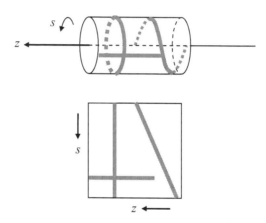

the surface of a right-circular cylinder unfolded into a rectangle; a straight line in the rectangle transforms into a line, circle, or helix on the cylinder, depending upon its orientation

    A photon travels with constant speed direction in a Euclidean space, such as a plane. A curved manifold is required to make it do otherwise. On a right-circular cylinder, a photon would travel in a straight line, circle, or a helix, depending upon which direction it is heading – so that the path becomes a straight line when the cylinder is unfolded into a plane. The circle corresponds to a photon at rest in the Euclidean dimension, but traveling the speed of light in the compact dimension with very high frequency. Such a photon would appear to have rest energy to a macroscopic being, if it could be detected, which is really relativistic energy from its kinetic energy in the compact dimension. The straight line path corresponds to a photon traveling directly along the Euclidean dimension – i.e. parallel to the axis of the cylinder (not on the axis itself, of course, since it is confined to the surface). This would behave as an ordinary photon, traveling the speed of light with the usual energy.

A photon traveling in a combination of the Euclidean and compact dimensions would wind around along a helix. Macroscopically, it would behave as a photon with effective rest-mass, traveling slower than the speed of light, though it actually travels the speed of light with zero rest-mass on the 2D manifold. The rotational energy it receives from winding around the compact dimension would effectively be interpreted as mass, mis-identified as rest-mass rather than relativistic mass at a macroscopic level since its component of velocity along the Euclidean dimension would be slower than the speed of light.

Any particle moving in a combination of the Euclidean and compact dimensions would be effectively identified as having more rest-mass than if it traveled directly along the Euclidean dimension. Again, this originates from the hidden rotational energy in the compact dimension. Conceptually, this simple classical model shows the origin of more massive states called *Kaluza-Klein excitations*. From a macroscopic point of view, any particle that can propagate into an extra dimension can behave as an ordinary particle or effectively as a slower-moving, more massive particle called a Kaluza-Klein excitation.

Any particle that travels in an extra compact dimension has an associated tower of heavier Kaluza-Klein excitations from an effectively 3D perspective.

The 4D cubinder can be cut like the right-circular cylinder, in this case unfolding into a 3D hyperplane. The possible paths of a photon – line, circle, or helix – are the same for both the cylinder and cubinder, and in both cases particles that propagate into the extra dimension have Kaluza-Klein excitations. The extra freedom in the Euclidean dimensions for the cubinder, compared to the cylinder, does not affect either of these features.

Let us consider communication between two point-particles. Again, for simplicity we will adopt the cylinder as our prototype, and the result will be the same for a 4D cubinder. These might be two electrons interacting by exchanging photons – signals of light – or two point-masses attracting one another through the exchange of gravitons.

Either way, each particle sends out messages radially, and they interact by receiving and sending messages.

Conceptually, it is easier to visualize the following scenario: Instead of two point-particles, imagine a tiny boat and lighthouse on the surface of the cylinder. The lighthouse radiates light outward, sending photons in every direction along the surface of the cylinder. Some of these photons will reach the boat. On a plane, only photons headed directly for the boat along the straight line that connects the boat and lighthouse would reach the boat, but on a cylinder, there is an infinite set of such photons. Some photons not heading directly for the boat along the shortest-distance route travel one or more times around the body of the cylinder and wind up reaching the boat. On a plane, only one lighthouse would be visible from the boat, but on the cylinder the boat sees an infinite set of lighthouses. The shortest-distance view results in the brightest lighthouse, while the other images are dimmer since light travels a greater distance along these routes. On the unfolded cylinder, these longer routes would consist of discontinuous jumps from one edge to another, but of course these edges meet when it is folded back into a cylinder.

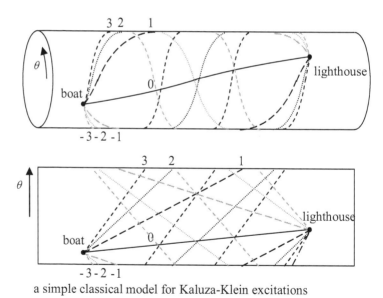

a simple classical model for Kaluza-Klein excitations

Conceptually, in this simple classical model, the shortest-distance route corresponds to the usual photon, while the longer routes correspond to the Kaluza-Klein excitations of the photon.

Macroscopically, these Kaluza-Klein excitations affect the flux of field lines for Gauss's law.

In order to understand the effect on Gauss's law, let us return to the 4D cubinder. A particle on the manifold of the 4D cubinder radiates field lines outward in four independent directions. According to Gauss's law, this 4D flux of field lines results in an inverse-cube force law. Two particles separated by a distance that is much smaller than the radius of the compact dimension experience this inverse-cube force.

Now imagine placing the two particles further away so that their separation is very large compared to the radius of the compact dimension. Geometrically, the flux of field lines still radiates outward in the three Euclidean dimensions, but it is fundamentally different along the compact dimension. In this case, the smallness of the compact dimension severely limits the flux of field lines. Since the flux radiates outward in three dimensions, with only a tiny contribution from the compact dimension, the particles effectively experience an inverse-square law. This is how compactification effectively restores the power of r observed for Newton's law of gravitation and Coulomb's law in our universe, despite extra degrees of freedom available for the flux of field lines in higher dimensions.

We can illustrate the different fluxes of field lines in the two extremes with the cylinder. Near a point-particle – the source of the field lines – on the surface of a cylinder, the field lines radiate outward two-dimensionally, like the spokes of a bicycle wheel, very similar to a point-particle in a plane. Far from the point-particle (i.e. compared to the radius of the cylinder), the field lines run mostly along the length of the cylinder. In the former case, they radiate along two independent directions, but in the latter case they primarily run along the Euclidean dimension. In this latter case, the flux of field lines is approximately one-dimensional.

Why do the field lines run mostly along the length of the cylinder far from the source? To see this, consider a test-particle far from the source. Regardless of its position in the compact dimension, it experiences a force of interaction from the messenger particle – the photon or graviton, depending upon whether the interaction is electromagnetic or gravitational – and also from each of the messenger's Kaluza-Klein excitations. For every Kaluza-Klein excitation that winds along a helical path in a clockwise sense, there is a counterpart that winds around in a counterclockwise sense. As a result, the components of force along the compact dimension largely cancel, while they all have a component in the same direction along the Euclidean dimension.

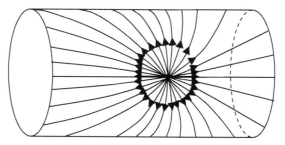

the field lines of a point source on the 2D surface of a cylinder appear 2D when viewed close to the point source (compared to the radius of the cylinder)

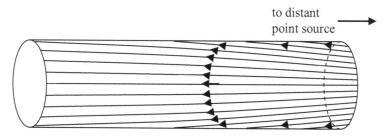

the field lines of a point source on the 2D surface of a cylinder appear roughly parallel when viewed far from the point source (compared to the radius of the cylinder), effectively reducing the dimensionality to 1D from a macroscopic perspective

8.5 Kaluza-Klein Excitations

We have seen a simplified classical picture for how Kaluza-Klein excitations come about. In particular, for a particle winding around the compact dimension in a helical path, the rotational kinetic energy that is hidden from a macroscopic point of view in the Euclidean dimension is effectively interpreted as rest mass. The particle appears to have a speed less than its actual higher-dimensional speed; this is its component of velocity along the Euclidean dimension. Particles appear to be effectively more massive and slower moving than they actually are.

We can improve upon this simplified classical model. Particles do not behave as classically thought: As we will see in Chapter 10, it is not possible to simultaneously determine the position and momentum (mass time velocity) of a particle, and particles exhibit both particle-like and wavelike properties, for example. A correct model of the behavior of particles in higher dimensions must account for the effects of quantum mechanics.

Here is how Kaluza-Klein excitations properly come about. We start by writing equations for the interactions of particles in the higher-dimensional theory. As much as possible, the higher-dimensional theory is constructed based upon straightforward generalizations of standard 3D formulations, as this is provides a viable method for the theory to agree with current experimental data, which we know presently agrees with the purely theory. Any measurements that we make are effectively 3D because the extra dimensions are hidden from our macroscopic vantage point, so we make an effective 3D theory from the higher-dimensional theory in order to compare with experiment. Mathematically, this is achieved through what is called a *Fourier series*. There series includes a zero-mode, which corresponds to the usual particle, plus heavier modes, which we interpret as the Kaluza-Klein excitations. The zero-mode represents the usual 3D physics, and the Kaluza-Klein excitations represent the effect of the compactified extra dimensions.

The interactions between particles are expressed with equations for their wave functions. A wave function for a particle can be used to determine various properties about the particle, such as the average value of its position or the uncertainty in its momentum over some time interval. The wave functions in the higher-dimensional theory involve Euclidean and compact coordinates. The Fourier series is a mathematically useful device for expressing a periodic function in terms of a series of sines and cosines. Since the particles experience periodic motion in the compact dimensions, it is useful to expand the wave functions for the particles with Fourier series. In doing so, the Euclidean component of the wave function can be separated from the compact component.

The Fourier series for the wave function for a particle includes a purely Euclidean term, which behaves as an ordinary particle, and a tower of terms involving sines and/or cosines of compact coordinates. Each term of this tower represents a particle that would behave very much like the ordinary particle – called a *Standard Model particle* –

except that it would have much more rest-mass. These are the associated Kaluza-Klein excitations of the particle.

The rest-masses of the Kaluza-Klein excitations are related to the rest-mass of the associated Standard Model particle. To good approximation, in the case of a single circular compact dimension, the lightest Kaluza-Klein excitation has a tree-level[6] rest-mass given by $m_1 = \sqrt{m_S^2 + m_K^2}$, where m_S is the rest-mass of the Standard Model particle, m_1 is the rest-mass of the lowest-lying Kaluza-Klein excitation of that particle, and m_K is inversely proportional to the radius of the extra dimension. The second lightest Kaluza-Klein excitation has a rest-mass of $m_2 = \sqrt{m_S^2 + 4m_K^2}$, the third lightest has a rest-mass of $m_3 = \sqrt{m_S^2 + 9m_K^2}$, and so on according to the pattern $m_n = \sqrt{m_S^2 + n^2 m_K^2}$.

Since the rest-mass of the Kaluza-Klein excitations all depend upon m_K and because m_K is inversely proportional to the compactification radius, a smaller compactification radius results in greater rest-masses for the Kaluza-Klein excitations. Naïvely, it might seem that it should be easier to detect such Kaluza-Klein excitations for *smaller* extra dimensions, as this would result in more massive Kaluza-Klein particles. However, we will learn in the next chapter that it is generally much easier to detect less massive particles than more massive ones. This means that it would be easier to discover Kaluza-Klein excitations if the extra dimensions are larger; this perspective should agree with your intuition.

8.6 Large Extra Dimensions

The extra dimensions predicted by string theory were originally expected to have a size on the order of the Planck length. Expressed in terms of three fundamental constants – the speed of light in vacuum ($c = 3.00 \times 10^8$ m/s), Planck's constant ($h = 6.63 \times 10^{-34}$ J·s), and the proportionality constant from Newton's law of universal gravitation

---

[6] There may also be loop-level contributions. What is meant by *tree* and *loop level* is deferred until Chapter 9.

($G = 6.67 \times 10^{-11}$ N·m²/kg²) – the Planck length equals $\sqrt{\dfrac{hG}{2\pi c^3}}$, which works out to about 10^{-35} m.

To see how incredibly tiny this is, examine a meterstick or some object that is about a meter long, such as a golf club. Imagine cutting the meterstick into a thousand tiny chunks of about the same size. Each chunk would be about a millimeter wide. Now take one of these chunks and split it widthwise – not lengthwise – a thousand times, dividing it into a thousand splinters. Each splinter would be about a micrometer wide, which is one millionth the size of the original meterstick. If you could split one of these splinters into a thousand more pieces – an idle application of nanotechnology – this would create very tiny pieces, just one nanometer wide. One nanometer is a billion times narrower than the length of the meterstick. Still, a nanometer is ten times larger than a typical atom. The nucleus of a typical atom, a mere 10^{-15} m in diameter, is one hundred thousand times smaller than the size of the atom, and 1,000,000,000,000,000 times shorter than the meterstick. Yet, the nucleus of a typical atom is incredibly enormous compared to the Planck length: It is 10^{20} times larger than the Planck length – an even greater difference in magnitude than the meterstick is compared to the nucleus. You would have to split the nucleus of an atom into 100,000,000,000,000,000,000 pieces in order to discover something the size of the Planck length.

Extra dimensions as small as the Planck length are so small that it is unfathomable that their presence could ever be detected. However, a few highly-regarded papers in the 1990's have motivated the case for extra dimensions to be much, much larger than the Planck scale. These are called *large extra dimensions*, even though they are smaller than a millimeter; the adjective *large* is used in comparison with the Planck scale.

The motivation for large extra dimensions stems from the observation that there are two fundamentally different energy scales in nature: the electroweak scale and the string scale. The electroweak scale is the energy scale associated with the electromagnetic and weak interactions between particles and the string scale is the fundamental energy scale for interactions between strings.

Similarly to how Maxwell's equations unify the seemingly different electric and magnetic fields into a single electromagnetic field in the sense that electric and magnetic phenomena are two different manifestations of one fundamental electromagnetic interaction, the Standard Model of elementary particles unifies the electromagnetic and weak nuclear forces into a single electroweak interaction. Particles that interact via the electromagnetic or weak interaction have rest-energies of up to 175 GeV – that's 1.75×10^{11} eV since the metric prefix giga (G) serves as a multiplier of one billion (10^9). The electron-volt (eV) is the amount of energy needed to accelerate an electron through a potential difference of one Volt (a Joule of energy divided by a Coulomb of charge). The as of yet undiscovered Higgs particle predicted by the Standard Model, which serves to give masses to the other particles, may have a somewhat larger mass, but less than a factor of ten at most. The supersymmetric partners of the Standard Model particles predicted by superstring theory are expected to be somewhat heavier than the Standard Model particles, but still on the TeV scale. The prefix tera (T) equals one trillion (10^{12}). The electromagnetic and weak nuclear forces are unified in the electroweak theory, which has been verified with high-precision experiments, with a notable exception being that we are awaiting the results of higher-energy experiments that have excellent prospects for discovering the Higgs particle.

    This fixes the electroweak scale at an energy of about 10^{12} eV. This equates to about 10^{-7} J, where the Joule (J) is the SI unit of energy. To get a feel for this, imagine jumping one meter into the air. A person weighing 200 pounds has a mass of about 100 kilograms (rounding to the nearest hundred) and a weight of about 1,000 Newtons in SI units. Jumping one meter into the air, gravity does 1,000 J of work to bring the person to a temporary halt at the top of the trajectory. The kinetic energy of the jump, 1,000 J, is ten billion times as much energy as the electroweak scale. But when you consider that the person who jumps has more than Avogadro's number (6.02×10^{23}) of particles, the electroweak scale is huge compared to the jumping energy per particle.

    The energy corresponding to the string scale is related to the size of the extra dimensions. In the previous section, we learned that smaller extra dimensions result in a greater mass for the Kaluza-Klein excitations, and more mass means more energy through Einstein's equation. For extra dimensions as small as the Planck length, the string scale is the Planck scale, which equals $\sqrt{\dfrac{hc^5}{2\pi G}}$. This is on the order of 10^{27} eV. Strings interact via a single, most fundamental, unifying interaction at this energy scale. This is an enormous amount of energy to be experienced in an interaction involving only a couple of elementary particles – well beyond what we could ever hope to produce in an experiment.

Here is the problem with the two different energy scales. One aesthetic motivation for superstring theory is that it is presently our best candidate for an ultimate theory of everything. Such a Grand Unified Theory would explain that the gravitational, electromagnetic, strong nuclear, and weak nuclear interactions are all just different manifestations of a single underlying interaction. But if there is just one underlying interaction, why are there two fundamental scales – electroweak and string – that differ so drastically in magnitude?

We shouldn't be surprised to find two scales that differ so much in magnitude. The main problem with developing a theory of everything is incorporating gravity into the theory. To see this, recall that the gravitational attraction between two electrons is about 10^{42} times weaker than their electromagnetic repulsion. Gravity is the most obvious force to intelligent macroscopic beings, yet it turns out to be extremely weak in comparison to the other three forces. We feel the gravitational pull of the earth because there are an astronomical number of molecules in the earth attracting the molecules of our body gravitationally; the gravitational attraction between just two particles is miniscule. The drastic difference in the two scales stems from the observation that the electromagnetic and gravitational interactions are dominant at totally opposite extremes.

Superstring theory is a theory of gravity in that it *can* unify the four fundamental forces – showing that it can do so in a way that is consistent with the observed universe and make plausibly testable predictions, though, is a theoretical challenge. Yet, superstring theory still features two drastically different scales – electroweak and gravity.

From an aesthetic point of view, this is an undesirable feature of the theory. That is, if it is the fundamental theory of everything, featuring one fundamental force, why isn't there a *single* fundamental scale? At the very least, if the ultimate theory of everything does feature two fundamentally different scales, it should be able to explain *why* there are two scales. From an experimental perspective, the underlying physics of quantum gravity can only be tested at the string scale, which is physically inaccessible in this picture.

This problem is not merely conceptual: Mathematically, an extremely delicate cancellation (one part in 10^{15}) is needed in supersymmetry (introduced in Chapter 10) in order for the theory to be plausible, which comes from the ratio of the two scales. This mathematical problem associated with two fundamental scales in nature is called the *hierarchy problem*. The details are much more technical and mathematical than suitable for our purposes here, but we can appreciate the delicate nature of the cancellation in terms of error analysis.

Sources of error are involved in *any* scientific measurement shy of counting a small number of objects. Errors are not mistakes, but inherent limitations. For example, if you wish to measure the volume of this book, you will encounter a limitation in the precision of the measuring device. You can purchase fancy equipment involving lasers and sensors to improve upon using a ruler, but you will still run into other problems. If you could view the book microscopically, you would see that the boundary is determined by particles that are moving, and eventually you would run into inherent errors from the uncertainty principle in quantum mechanics. You *can't* measure length, time, mass, and other physical quantities (except for number) exactly – there are inherent limitations called *errors*.

Let us quantify some common errors. We can measure gravitational acceleration near the surface of the earth with a stopwatch and meterstick. In this case, it should be quite reasonable to measure it to within 1 part in 10; some good technique and equipment would be necessary to claim 1 part in 100. If you manufacture microwave ovens, you will be concerned with parts per thousand – i.e. how many microwaves are defective for every thousand sold. If you drink a bottle of water, you should be concerned with parts per million; if there is a contaminant, this fraction is very important to your health. It is quite an achievement to purify water so that only a few parts per million are not H_2O molecules – imagine sorting through a bottle of one million pennies by hand to eliminate all of the Canadian pennies. In comparison, supersymmetry requires a cancellation between terms of 1 part in 1,000,000,000,000,000. This delicate cancellation is called *fine-tuning*.

The traditional approach toward developing a Grand Unified Theory is as follows. The interactions between vibrating strings occur at the Planck scale, as described by an underlying, fundamental theory unifying quantum gravity and the other forces. The theory in its purest form applies at the extremely high energies of the Planck scale. At much lower energies – namely, the electroweak scale – an effective field theory governs the interactions between elementary particles. At the electroweak scale, much different manifestations of the single underlying interaction appear as the four fundamental forces. Some mechanism, such as supersymmetry breaking, is responsible for creating the lower-energy effective field theory.

This traditional approach maintains the two fundamental scales observed in nature, and aims to explain the origin of the hierarchy. Although, in this picture, there is no hope for confirming the fundamental theory of quantum gravity at the Planck scale, current and upcoming experiments in high-energy physics are expected to discover *how* nature breaks the true theory into an effective theory at the electroweak scale and to provide low-energy clues to help us piece together the puzzle of higher-energy gravity.

Here is a summary of a few key problems with the conventional Grand Unified Theories. There is a large hierarchy problem with two vastly different fundamental energy scales – the electroweak scale and Planck scale. This is largely aesthetic – i.e. why should the ultimate theory of nature consist of two disparate fundamental scales? Along with this is a mathematical fine-tuning problem that requires extremely delicate cancellations. The true theory of gravity seems largely philosophical in this picture, as there is no hope of performing experiments at the Planck scale; the best we can hope for are a few clues from experiments conducted at the electroweak scale.

The proposal of large extra dimensions [T2-T3] offers a novel solution to this hierarchy problem. The main idea comes from realizing that the electroweak theory has been observed at the electroweak scale, whereas gravity has not been tested even remotely close to the originally predicted Planck scale. Newton's law of gravitation has only been tested to within about a tenth of a millimeter, whereas the Planck length is 10^{-35} m. A major assumption of the traditional approach is that Newton's law of gravity extrapolates over 30 powers of 10, but there is no experimental evidence to show that it in fact has the same form at distances under about a tenth of a millimeter. So it is possible that the true Planck scale is really closer to 10^{12} eV instead of 10^{27} eV. That is, maybe there really is just one fundamental scale, at 10^{12} eV, which is the observed electroweak scale.

It has been shown [T2-T3] that superstring theory can feature a single fundamental scale, at a scale much lower than 10^{27} eV, if the extra dimensions are much larger than originally predicted. That is, the Planck scale is not a fundamental scale of nature. Gravity and the electroweak gauge interactions are unified at a single fundamental scale, which is normally called the elctroweak scale. Because gravity is so incredibly weak (i.e. the gravitational attraction between two electrons is about 10^{42} times weaker thean their electrical repulsion), it was originally expected that gravity would be strong only at a very high energy scale – namely, the Planck scale (10^{27} eV). The Planck scale and Planck length are inversely related: The extra dimensions originally predicted by superstring theory were thought to be so minuscule (about 10^{-35} m in size) because the Planck scale is so large. However, it turns out that gravity can be strong at the electroweak scale (10^{12} eV) – i.e. unified with the electroweak interactions – if the extra dimensions are very large (as much as a fraction of a millimeter). Larger extra dimensions means a greater flux of gravitational field lines in the extra dimensions, which significantly affects the power of r in Newton's law of gravity (as explained in Chapter 7); hence, the reciprocal relationship between the size of the extra dimensions and the corresponding energy scale.

This proposal of large extra dimensions inherently solves the associated fine-tuning problem, removes the aesthetic problem with two disparate fundamental energy scales, and comes with an added bonus: The extra dimensions predicted by superstring theory are now motivated to be much, much larger than originally thought – as large as a fraction of a millimeter (depending upon the model, as we will see in the following chapter). That is, in this framework, it is very plausible for current and upcoming experiments to detect the presence of superstring theory and large extra dimensions in our universe. This elevates superstring theory from a rather philosophical, aesthetically pleasing theory that can at best provide a little very indirect evidence to a scientific theory that can more plausibly be physically tested in the foreseeable (and likely near) future. In consequence, this novel solution to the hierarchy problem has stimulated an abundance of professional papers on the theoretical and phenomenological ramifications of superstring-inspired models with large extra dimensions.

8.7 Higher-Dimensional Models

Henceforth, we shall consider models with one or more large extra dimensions. Since these large extra dimensions are superstring-inspired and since superstring theory predicts six extra dimensions,[7] the model should accommodate six extra dimensions. (There are ten dimensions of spacetime in superstring theory. Three of these are the macroscopic spatial dimensions that we observe and one dimension must correspond to time, which leaves six extra dimensions.) However, all six extra dimensions need not be the same size, and all need not be large. So an open question in developing a higher-

---

[7] We will consider M theory briefly in Chapter 10, where there is effectively another dimension.

dimensional model is how many large extra dimensions there are and how large they are.

Another important issue is which particles propagate into which extra dimensions. Particles that we observe in our universe evidently have access to the three usual dimensions of space, but it is not necessarily the case that all particles can propagate into all of the dimensions. The only particle that must propagate into the extra dimensions is the graviton – the mediator of the gravitational interaction between two masses – since string theory is a theory of gravity. Other particles may propagate into just a subspace of the full $(9+1)$-D spacetime.

One common model is that only gravitons propagate into the extra dimensions, and all of the other particles reside in the usual three dimensions. This ADD model, so named in honor of three authors – Arkani-Hamed, Dimopoulos, and Dvali – whose paper [T2] has stimulated much of the current research on large extra dimensions, naturally explains why we are not able to "see" extra dimensions in our universe. For one, the photons – the particles of light that we see – are unable to propagate in the extra dimensions, and for another, the particles that we are made of – namely, protons, neutrons, and electrons – can not move along these extra dimensions either. The effect that the presence of such large extra dimensions has on our experience must be probed through gravitational interactions. We say that the usual particles reside on a D_3-brane (a 3D subspace of the higher-dimensional theory).

We will refer to the ADD model as a scenario in which only the gravitons propagate into any extra dimensions. All other particles are confined to the 3D Standard Model wall.

However, it is also possible that all of the particles can propagate into one or more extra dimensions. This is termed the universal extra dimensions (UED) model [T4]. The number of universal extra dimensions δ can be between one and six. In this case, the usual particles reside on a $(3+\delta)$-D subspace, called a $D_{3+\delta}$-brane, while the gravitons propagate into all of the extra dimensions (which may be a greater number than δ – one or more of the extra dimensions may be universal, but not necessarily all of them). Universal extra dimensions cannot be as large as extra dimensions accessible only to gravity in order to explain why we are unable to "see" any such extra dimensions. We will examine the exact bounds in the following chapter; it turns out that they must be as small as about the size of an atomic nucleus, which reconciles why we are unable to notice any motion in such extra dimensions.

All of the particles propagate into one or more extra dimensions in the UED model. In this case, the Standard Model wall is a 4D or higher subspace; only the gravitons propagate into the bulk – extra dimensions beyond the Standard Model wall.

Finally, it is also possible to have extra dimensions accessible to some of the usual particles, but not all of them. For example, particles like protons, neutrons, and electrons may reside in the usual 3D space, while photons and the other gauge bosons – i.e. the mediators of the four fundamental forces – may propagate into one or more extra dimensions.

Which particles do or do not propagate into extra dimensions has a significant effect on the bounds of large extra dimensions, as does the size and shape of the extra dimensions.

Further Reading:

The original research papers that motivated large extra dimensions include work by Arkani-Hamed, Dimopoulos, and Dvali [T2], along with work by Antoniadis [T3]. See [T4] for the proposal of the UED model. These papers are highly technical; for a much more accessible article on this topic, see [A1]. For accessible references on string theory, including compactification, see the suggested readings for Chapter 10.

9 Experimental Searches

Here we explore current and upcoming experimental searches for the presence of large extra dimensions in our universe. These will not be direct observations – i.e. we will not see the extra dimension itself – but these experiments will either provide compelling indirect evidence for extra dimensions or place bounds on the size of any extra dimensions. Large extra dimensions can be detected indirectly through one of two means: an increase in gravitational flux that impacts the power of r in Newton's law of gravitation (as described in Chapter 7) or the production of KK excitations (described in Chapter 8). As we will see, how the results of the experiments will be interpreted will depend somewhat on the model – e.g. there will be a significant difference between the ADD and UED models (defined at the end of Chapter 8).

9.0 Tests of Newton's Law of Gravitation

As explained in detail in Chapter 7, extra dimensions in our universe would increase the flux of gravitational field lines, thereby affecting the power of r in Newton's law of gravitation. The larger the size of the extra dimensions, the greater the effect that extra dimensions have on the power of r. Direct tests of Newton's law of gravitation aim to measure the power of r very precisely to look for possible deviations – i.e. to see if the power of r might be slightly greater than 2. A significant deviation from 2 would be a clear indication of new physics, for which large extra dimensions would offer an explanation. In this case, additional experimental confirmation, such as the detection of KK excitations (described in following sections), could show whether or not the source of new physics is, in fact, due to the presence of large extra dimensions in our universe. On the other hand, if the power of r measured is consistent with 2 within the experimental uncertainty, this does not rule out the possibility of extra dimensions; rather, the uncertainty would provide a bound for the size of any extra dimensions. In this case, the smaller the uncertainty, the more stringent the bound – where the bound is the maximum size that the extra dimensions can have and still be consistent with experimental observations. Smaller extra dimensions would have a smaller effect on the power of r.

Naïvely, what we would like to do is separate the centers of two masses m_1 and m_2 by a distance r, measure the magnitude of the gravitational force F of attraction between them, set the result equal to $G\frac{m_1 m_2}{r^{2+\alpha}}$, and compute α. Also, the uncertainty σ_F in the measured force F can be used to establish the uncertainty σ_α in the quantity α. If we find that $\alpha = 0.020$ and $\sigma_\alpha = \pm 0.005$, for example, then this 4σ deviation (meaning that the results differ by 4 sigma, or 4 uncertainties) is a clear discrepancy – an indication of some new physics mechanism. As a counterexample, if $\alpha = 0.010$ and $\sigma_\alpha = \pm 0.015$, then the results agree within 1 sigma, since $2+\alpha$ lies in the range $2.010 - 0.015 < 2+\alpha < 2.010 + 0.015$; in this case, the uncertainty $\sigma_\alpha = \pm 0.015$ would be used to establish the maximum size that extra dimensions could have and not cause the power of r to deviate from 2 by more than 0.015.

The problem with this is that it easy to measure the force between two masses when one (or both) of the masses is astronomical, but the gravitational forces between two objects that have masses of about a kilogram is extremely tiny. This follows since the gravitational constant G is very tiny in SI units (6.67×10^{-11} Nm^2/kg^2). So two 1-kg masses placed 1 m apart experience an attractive gravitational force of just 6.67×10^{-11} N. (Note that 1 pound equates to 4.45 N.) This is an extremely tiny force to measure. We can make the force larger by using larger masses – e.g. loading two semi trucks up so that each has a mass of, say, 10,000 kg. Even so, the force of attraction between the semi trucks is a mere 6.67×10^{-3} N – less than one-hundredth of a Newton, about a thousandth of a pound. Certainly, not enough to overcome resistive forces such as friction between the tires and ground or air resistance; such a force would not be observable. Alternatively, we could produce larger forces by making r smaller – i.e. placing the two objects closer together. This requires very dense objects, with lead providing the limit to what density we may expect to achieve using elements from the periodic table, since there is an inherent limitation on r. You can't make r less than 1 m in the case of two semi trucks because r is the distance between their centers, not their edges.

Furthermore, the extra dimensions have a greater effect if r is small and a smaller effect if r is large. If we measure the force of attraction between the earth and moon, the force is very large 1.99×10^{20} N, but in this case an extra dimension smaller than a millimeter would have no observable effect since the two objects are over a hundred thousand kilometers apart. A classical graviton that winds its way along a helical path in an extra dimension that is about a millimeter wide, but about a hundred thousand kilometers long, travels only very minutely further than a graviton that travels in a perfectly straight line from the earth to the moon – unless it winds around numerous times, in which case this corresponds to a much heavier Kaluza-Klein excitation, which has a less significant effect (and it cancels the component of force from a corresponding graviton winding around the extra dimension in the opposite rotational sense). However, for two objects that are a millimeter apart (i.e. their centers are separated by 1 mm), an extra dimension as large as a millimeter would have a very large effect on the flux of gravitons between the two objects.

So if we want to discover extra dimensions that are smaller than a millimeter, we want the separation between the centers of the two objects to be very small – say, a millimeter or smaller. However, this makes it much more difficult to measure the gravitational force of attraction.

The power of r in Newton's law of gravitation has been confirmed to be very close to 2, with extremely little uncertainty, for astronomical distances because astronomical forces and their effects on objects observable in the night sky are rather easy to measure. On the other hand, Newton's law of gravitation has not been confirmed very well – meaning that the power of r has much more uncertainty – at very short distances, especially in the sub-millimeter regime. The current upper bound on the size of extra dimensions is about a tenth of a millimeter.

This bound of about a tenth of a millimeter corresponds to the ADD model. It must be much smaller in the case of the UED model, of course, since extra dimensions would be quite obvious if ordinary particles could move in an extra dimension as large as a millimeter or so. We will see in later sections that UED extra dimensions must be as small as about the size of an atomic nucleus. Even if there are UED extra dimensions, it may very well be the case that only one or more are UED extra dimensions, while the others are accessible only to gravitons. The largest extra dimensions in such a UED model would have a gravitational phenomenology similar to the ADD model.

Let us consider some of the fundamentals of the theory that we would like to test. As discussed in Chapter 7, for δ extra dimensions, Newton's law of gravitation has the form $G_\delta \frac{m_1 m_2}{r^{2+\delta}}$ if the separation r between the two masses is very small compared to the radius of the extra dimensions R, where we are placing the subscript δ on the gravitational constant G_δ in order to distinguish it from the effective

(3+1)-D gravitational constant G (since it turns out that they can not have the same value). In the other extreme – i.e. where the separation r between the two masses is very large compared to the radius of the extra dimensions R, it can be shown that Newton's law of gravitation has the approximate form $G_\delta \frac{m_1 m_2}{R^\delta r^{2+\alpha}}$, where α is a very small correction in this limit.

As we have seen, we can not hope to make r nearly as small as R in the foreseeable future, so any measurement that we make will correspond to the latter case, $G_\delta \frac{m_1 m_2}{R^\delta r^{2+\alpha}}$. This must be very nearly equal to $G \frac{m_1 m_2}{r^2}$, since the usual form of Newton's law of gravitation, $G \frac{m_1 m_2}{r^2}$, is observed to hold for experiments where r is about a millimeter or greater in size. As we develop experiments to probe smaller and smaller r, we will either discover that α does not equal zero or place smaller and smaller limits on how large α can be.

Since we know that α must be fairly small – at least until we develop experiments where r becomes microscopic – let us ignore the small effect that α has for a moment. In doing so, we see that $\frac{G_\delta}{R^\delta}$ is approximately equal to G, or $G_\delta = GR^\delta$. Recall that R is the size of the extra dimensions, not to be confused with the separation between the two masses r; we have in mind experiments where r is much larger than R, since we are limited in how small we can make r and still measure the attractive force precisely, while we expect the radius R of any extra dimensions to be quite small. Also, recall that δ is the number of extra dimensions. Finally, recall that the Planck scale was defined as $\Lambda_P = \sqrt{\frac{hc^5}{2\pi G}}$, or put another way, G is proportional to $1/\Lambda_P^2$. We can use this to write the relation $G_\delta = GR^\delta$ in terms of the Planck scale as $\Lambda_P^2 = R^\delta \Lambda_{P,ed}^{2+\delta}$, where $\Lambda_{P,ed}$ is the higher-dimensional Planck scale and Λ_P is the traditional value of 10^{27} eV.

In the framework of large extra dimensions, $\Lambda_{P,ed}$ is proportional to the electroweak scale – which is about 10^{12} eV – by some fundamental constants while $\Lambda_P \sim 10^{27}$ eV is the effective

(3+1)-D Planck scale that we effectively observe (since we are macroscopically "ignorant" of any extra dimensions that may be present in our universe). Solving for the radius of the extra dimensions R and carefully accounting for the proportionality between the electroweak scale and $\Lambda_{P,ed}$, it turns out that R is about 10^{11} m (one hundred trillion meters) if $\delta = 1$, between a micrometer and 0.1 mm if $\delta = 2$, and progressively smaller for larger numbers of extra dimensions δ.

The case $\delta = 1$ is clearly ruled out, as it would require a drastic variation of Newton's law of gravitation at astronomical distances – in sharp contrast with experimental observations. That's okay, though, since superstring theory predicts 6 extra dimensions. The case $\delta = 2$ is particularly interesting, as it lies right on the verge of potential experimental discovery. Although superstring theory predicts 6 extra dimensions, the case $\delta = 2$ is still viable in the case of asymmetric extra dimensions – i.e. where there are two large extra dimensions and four comparatively smaller ones; similarly, there can be $\delta = 1$ large extra dimension and five smaller ones.

Let us return to the experimental details. We saw the difficulty in placing two dense masses close to one another to test Newton's law of gravitation at short range. Experiments that have the potential to produce precision measurements of Newton's law of gravitation at very short distances must either overcome this difficulty or apply a different experimental technique.

Newton's law of gravitation was first measured at short-range by Henry Cavendish in 1798. This historic experiment led to the first calculation of the gravitational constant. Cavendish used four dense spheres, instead of just two – i.e. he did not simply place two dense spheres a fixed distance apart and attempt to measure the attractive force. Rather, he used two large spheres and two small spheres. He connected the two small spheres by a thin, lightweight rod and suspended them from a vertical wire. He placed one large sphere near each of the smaller spheres – a distance r separated the large sphere from the small sphere. Each pair of spheres was far from the other compared to r so that the dominant attractive force was between larger and small spheres. The rod that connected the small spheres was not aligned with the line connecting the two large spheres, which resulted in a torque that caused the rod to rotate. Since the attractive force was quite small, Cavendish placed a mirror on the vertical wire. When the rod rotated, so did the mirror. Careful observation of an image reflected by the mirror allows a precise determination of the angle of rotation, which ultimately led to a precise measurement of the attractive force between the small and large masses. Even applying this clever technique with modern-day equipment, it is quite challenging to probe the sub-millimeter regime.

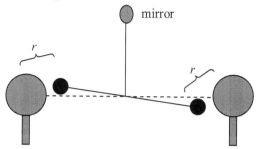

the Cavendish apparatus for measuring gravitational attraction

Aside from the difficulties that we have already seen in achieving an appreciable force between two dense masses placed a short distance apart, there is another significant problem: At smaller and smaller distances, the other fundamental forces of nature dominate over gravity. It is very difficult to measure a gravitational force of attraction when there is a much stronger Coulomb force of attraction or repulsion between the charged particles that comprise the masses. Even though the two masses may be neutral overall, they are made of charged particles, and the distribution of charges actually leads to significant electrical force between the masses – since the electromagnetic force is so large compared to gravity. As an example, consider rubbing a balloon against your pant leg and sticking it to a wall. The balloon is electrically charged, as it exchanges electrons with your pant leg, but the wall is neutral. So why does the balloon adhere to the wall? Charged objects attract or repel other charged objects – they don't attract to neutral objects, right? Well, the molecules on the wall, although neutral, have their charges distributed like electric dipoles – i.e. effectively with positive charge at one end and negative charge at the other. In the presence of the balloon, these polar molecules align – one end being attracted to the charged balloon and the other repelled by it. The balloon then adheres to the wall through its attraction to the ends of the polar molecules closest to it. Such interactions between charged objects and polar molecules – as well as polar-polar interactions, polar-nonpolar interactions, and nonpolar-nonpolar interactions – can be large compared to gravitational forces, which adds to the difficulty in testing Newton's law of gravitation at short-distance scales.

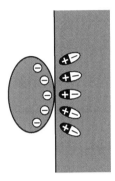

a charged object attracting to a neutral object

Despite these challenges, current and upcoming experiments have been designed to measure the attractive force from Newton's law of gravitation more precisely [T5] – i.e. to determine the power of r very precisely for separations of less than a millimeter. One of the techniques that is presently successful involves the Casimir effect. The Casimir force is yet another force that two neutral metallic plates exert on one another when separated by a vacuum. Accounting for the Casimir force, gravitational attraction, and van der Waals forces – similar to the types of interactions between neutral objects mentioned above – this is presently among our best techniques for determining α in the sub-millimeter regime.

As experimental techniques lead to more and more precise measurements of α in Newton's law of gravitation, we come closer to detecting the presence of large extra dimensions in our universe – and if we do not detect them, then we are able to place more stringent bounds on how large they can be. Direct tests of Newton's law of gravitation are just one means of probing for extra dimensions. There are also collider searches, astrophysical and cosmological searches, and various other methods underway. These searches involve the interactions between elementary particles. Thus, we will take an interlude in the next section to understand elementary particles and how they interact and then we will return to experimental searches for extra dimensions in the remainder of this chapter.

9.1 Elementary Particles and Their Interactions

Ordinary matter – i.e. macroscopic material objects with which we are familiar – is composed of atoms consisting of protons and neutrons in the nucleus surrounded by electron clouds. The protons and neutrons themselves are composed of fractionally-charged particles called quarks. Specifically, a proton is a bound state of two up quarks and one down quark, while the neutron is a bound state of two down quarks and one up quark. An up quark has a charge of $+2e/3$ and a down quark has a charge of $-e/3$, where e represents the charge of a proton (so $-e$ corresponds to the charge of an electron). It is easy to check that the charge of two up quarks and one down quark equals the charge of a proton and, similarly, that the charge of two down quarks is canceled by the charge of the up quark in a neutron. Quarks are never observed to exist as free particles, but only form states with two or three quarks bound together as a single object, like the proton and neutron. The up quark and down quark have nearly equal masses, which explains why protons and neutrons have nearly equal masses. Since the neutron is slightly heavier than the proton, it follows that the down quark must be slightly heavier than the up quark. Up and down quarks, however, are much heavier than electrons. Quarks and electrons are evidently elementary particles – i.e. not made up of yet smaller constituents.

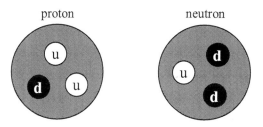

protons and neutrons are composed of fractionally-charged elementary particles called quarks

Ordinary matter is composed of three different types of elementary particles: up quarks, down quarks, and electrons. Yet several other elementary particles have been observed that are not components of ordinary matter. These other elementary particles can be produced by high-energy cosmic rays traveling through space upon interacting with particles in the earth's upper atmosphere, processes that occur in stars, and high-energy collider experiments where particles are smashed together, for example.

There are two particles, called the muon μ^- and tau lepton τ^-, which behave very similarly to electrons except that they have much more mass than an electron. The muon and tau lepton both have negative charge equal to that of the electron and have the same spin as an electron. Muons and tau leptons also have very similar interactions with gauge bosons, such as the photon. Ordinary atoms are surrounded by electron clouds, though, and not muons nor tau leptons because electrons are stable, while the muon and tau lepton are not. Since the muon is much heavier than the electron, it can decay to an electron and other particles in a tiny fraction of a second; similarly for the tau lepton, except that it has more decay options since it is much heavier – e.g. it can first decay to a muon plus other particles. So, when muons and tau leptons are produced in our universe, they only last a very brief time. Electrons cannot decay to lighter particles, though, so they are stable. This is why we see electrons in ordinary matter, but not muons or tau leptons. Electrons, muons, and tau leptons are collectively referred to as charged leptons. We say that there are three generations of leptons, referring to the electron, muon, and tau lepton.

Similarly, there are three generations of quarks. There are two quarks in each generation – i.e. there are three generations of charge $+2e/3$ quarks and three generations of charge $-e/3$ quarks, called up-type quarks and down-type quarks, respectively. The up and down quarks form the lightest generation, the second generation consists of the charm and strange quarks, and the third generation includes the top and bottom quarks. The up, charm, and top have positive charge, while the down, strange, and bottom have negative charge. The top quark is the heaviest quark, and the bottom is the heaviest down-type quark.

There are actually three generations of pairs of leptons, too. In addition to the charged leptons, there are neutral leptons called neutrinos. These include the electron neutrino, muon neutrino, and tau neutrino.

The particles are classified according to their properties and interactions with other particles. The quarks are the only particles that interact via the strong nuclear interaction; this is the interaction that binds quarks together in bound states, and binds protons and neutrons together in the nucleus. The strong nuclear interaction is mediated by the gluon; so named from the sense that it is the "glue" that holds nuclei together.

The electromagnetic interaction is mediated by the photon; this is how charged particles communicate. Any two particles that have charge experience electromagnetic interactions. This includes the quarks and charged leptons. Protons repel electrically, and electrons are attracted to the nucleus, through the electromagnetic interaction. Photons can be absorbed or emitted by electrons undergoing transitions between energy levels in an atom via the electromagnetic interaction; such transitions are seen in atomic spectra.

The weak nuclear interaction is mediated by $W^\pm$ and Z bosons. All quarks, charged leptons, and neutrinos can interact with these gauge bosons. The weak nuclear interaction is seen when a neutron decays to a proton plus an electron and electron anti-neutrino, which is a fundamental process to radioactive nuclear decays. Whereas the other gauge bosons – i.e. the photon, gluon, and graviton – have zero rest-mass, the $W^\pm$ and Z bosons have rest masses that are 80 to 90 times heavier than a proton.

The up and down quark form a left-handed doublet in the sense that left-chiral up and down quarks can couple to a $W^\pm$ boson; the charm and strange as well as the top and bottom form analogous left-handed doublets. Similarly, the left-chiral electron and electron neutrino can couple to a $W^\pm$ boson, and so form a left-handed doublet; the muon and muon neutrino as well as the tau lepton and tau neutrino also form left-handed doublets. This explains the grouping of elementary particles into three generations of quark and lepton pairs. Strangely, the right-chiral fields do not couple to the $W^\pm$ boson, and so are termed right-handed singlets. The grouping of elementary particles is summarized in the table below.

For every particle, there also exists an antiparticle. The antiparticle has the same mass as the particle, and its properties and behavior are the same as the particle except for having its quantum numbers reversed. One such reversal is its electric charge. The positron, for example, is the antiparticle to the electron. The positron is just like an electron, but with positive charge. Ordinary matter consists of protons, neutrons, and electrons, though, and positrons are extremely scarce. Positrons are produced in everyday interactions, but with an abundance of electrons around, positrons quickly interact with electrons; a large percentage of the interactions between electrons and positrons results in the production of two photons – called pair annihilation, as two particles with nonzero rest-mass are annihilated to produce two particles with zero rest-mass. As a result, when positrons are produced, they disappear very quickly after one or more interactions with electrons. Protons, neutrons, and all other particles have antiparticles, except that a couple of neutral particles serve as their own antiparticles (like the photon).

Volume 2: The Physics of the Fourth and Higher Dimensions...

	particle	charge	mass	spin	forces
	Quarks				
1	up quark	$2e/3$	$\sim 3\times 10^{-3} m_p$	f	SN,EM,WN,G
1	down quark	$-e/3$	$\sim 5\times 10^{-3} m_p$	f	SN,EM,WN,G
2	charm quark	$2e/3$	$1.4 m_p$	f	SN,EM,WN,G
2	strange quark	$-e/3$	$0.11 m_p$	f	SN,EM,WN,G
3	top quark	$2e/3$	$183 m_p$	f	SN,EM,WN,G
3	bottom quark	$-e/3$	$4.8 m_p$	f	SN,EM,WN,G
	Leptons				
1	electron	$-e$	$5\times 10^{-3} m_p$	f	EM,WN,G
1	electron neutrino	0	$<10^{-8} m_p$	f	WN,G
2	muon	$-e$	$0.11 m_p$	f	EM,WN,G
2	muon neutrino	0	$<10^{-3} m_p$	f	WN,G
3	tau lepton	$-e$	$1.9 m_p$	f	EM,WN,G
3	tau neutrino	0	$<0.2 m_p$	f	WN,G

a table of elementary particles

e = charge of one proton, m_p = mass of one proton, f = fermion, SN = strong nuclear, EM = electromagnetic, WN = weak nuclear, G = gravity

particle	charge	mass	spin	source
Mediators				
gluons	0	0	b	color
photon	0	0	b	electric charge
$W^\pm$, Z	$\pm e$, 0	$86 m_p$, $97 m_p$	b	weak isospin
gravitons	0	0	b	mass

a table of the mediators of the fundamental forces (b = boson)

The Standard Model – the present minimal theory to explain the observed interactions between the elementary particles – requires one more particle, called the Higgs boson, in order to explain how the other particles are able to have rest-masses. The Higgs mechanism – i.e. the means by which the Higgs gives rest-masses to the other particles – is a fundamental component to the current theory for electroweak unification – i.e. it explains how the electromagnetic and weak nuclear interaction are two different manifestations of a single underlying interaction, which result from electroweak symmetry breaking. The success that this electroweak theory has enjoyed thus far is compelling for grander schemes of unification – with the ultimate goal of unifying all four fundamental interactions. The best candidate for unifying gravity with the other interactions is superstring theory, and, as we have seen, the case for large extra dimensions has been made to unify the four interactions at a single scale.

The Higgs and graviton have yet to be detected experimentally; there is strong indirect evidence to confirm the presence of quarks and their interactions with gluons in protons and neutrons; and the remaining elementary particles have been observed (more or less) directly. The Standard Model is very successful in terms of the amazing level of agreement between its theoretical predictions and experimental observations for processes that involve elementary particles [T7]. Although we await the discovery of the Higgs boson in the very near future – or a historic surprise from its non-discovery – the electroweak theory which features the Higgs has correctly explained many features of the Standard Model and had prior predictions confirmed. For example, it correctly predicts that three of the gauge bosons – namely, the $W^{\pm}$ and Z bosons – should have rest-mass, and indeed their masses have been measured and agree with the relationship that they are expected to have with other parameters in the electroweak theory.

However, there are many aesthetic features that the Standard Model does not explain, which provides much of the motivation for a simpler, more fundamental underlying theory – a Grand Unified Theory. As superstring theory is presently our best candidate for a fundamental theory, this serves as part of the aesthetic motivation for the presence of extra dimensions in our universe. Some of the issues that the Standard Model does not resolve include:

- Why are there three generations of matter? Are the heavier generations different manifestations of the lightest generation?
- Why is there a large number of elementary particles? Are these different manifestations of a smaller number of truly elementary particles? Is there a simple way to explain the wide range of rest-masses for neutrinos, charged leptons, and quarks?
- Can gravity be unified with the other interactions? Are gravity, electromagnetism, and the strong and weak nuclear forces different manifestations of a single underlying force?

This last issue is related to the two vastly different scales observed in nature – i.e. the gravitational and electroweak scale – discussed in the previous chapter. Thus, we can see how superstring theory with large extra dimensions provides some hope of resolving some of these aesthetic puzzles.

The presence of large extra dimensions in our universe would affect the interactions between elementary particles and also result in additional particles – i.e. the KK excitations described in Chapter 8. Searching for these additional particles or the indirect effects that large extra dimensions would have on the interactions between ordinary particles involves understanding high-energy physics phenomenology – namely, how to produce and detect elementary particles. In the laboratory, elementary particles can be produced and smashed together in high-energy colliders. Elementary particles and their interactions can also be studied in high-energy cosmic rays that reach earth's atmosphere, and through astrophysical and cosmological observations of the universe.

9.2 High-Energy Collider Phenomenology

At high-energy colliders, elementary particles are accelerated to very high speeds and "smashed together" – i.e. they interact with one another at very close range in high-energy collisions. This is the particle physicist's controlled laboratory. The alternative is to observe interactions between elementary particles that occur naturally, such looking for showers of particles produced by high-energy cosmic rays in earth's atmosphere or observing products from chemical reactions in stars. In the collider, physicists know where to place the detectors and when to make the measurements; in astronomical observations, astrophysicists also know to orient detectors toward stars or supernovae, but cannot control what reactions will occur when (though we may develop a good statistical expectation); and in high-energy cosmic rays, events are much less predictable. However, we are much more limited in how much energy is involved in terrestrial collider experiments, so while we have less control over astrophysical and cosmic ray reactions, these natural phenomena can probe much higher energies than we can conceivably hope of producing in the laboratory.

The first step in making a high-energy collider is to create beams of elementary particles. Beams of charged particles are both easy to produce and easy to control. It is particularly easy to make beams of protons and electrons. Hydrogen gas, which is rather abundant, consists of diatomic hydrogen molecules – each molecule consisting of two hydrogen elements bound together. The most common isotope of hydrogen has one proton and one electron, but no neutrons. The electrons can be stripped off of the nuclei (which in this case are just protons) through a process called ionization. This is fairly easy since the electrons have very little mass (inertia), and thus are easily accelerated; give the bound electron more energy than its binding energy and it becomes free of its attraction to the proton.

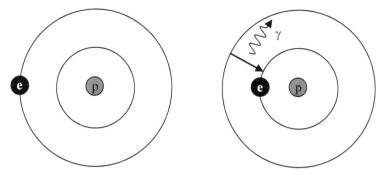

an electron in an excited state drops down to a lower-energy level, emitting a photon in the process in order to conserve energy

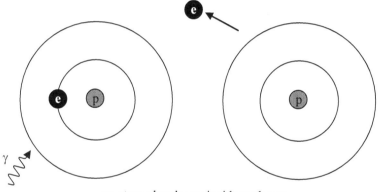

an atom absorbs an incident photon with enough energy to release the electron (i.e. to ionize the atom)

Electrons can become ionized simply by shining high-frequency electromagnetic radiation (i.e. light) on the hydrogen gas. A photon carries energy that is proportional to its frequency: $E = hf$, where h is Planck's constant, 6.626×10^{-34} J·s (the equation and constant come from Planck's novel solution to the ultraviolet catastrophe pertaining to blackbody radiation, and the full conceptual understanding of the formula came when Einstein applied it explain the photoelectric effect). The equation shows that higher-frequency electromagnetic waves, such as ultraviolet radiation and x-rays, consist of higher-energy photons. Such photons carry enough energy to overcome the electron's binding energy and ionize hydrogen atoms. The result is the separation of the electrons and protons that make up hydrogen atoms. This is just one way to supply the energy needed to ionize atoms.

Once separated, the charged particles (either electrons or protons) can easily be collected by attracting them toward an oppositely charged plate. Electrons will be attracted toward positively charged plates, while protons will be attracted toward negatively charged plates. Charging plates is easy – a battery or power supply connected to two separate pieces of metallic conductors will result in one conductor becoming positively charged and the other becoming negatively charged. However, the charged plates do not serve as collection plates in the ordinary sense – i.e. the idea is not to have the electrons or protons collect on a plate. After all, the plate itself consists of atoms that have protons, neutrons, and electrons, so it would not be useful to separate the protons and electrons from hydrogen gas to have them mix with other particles in a "collection plate." Rather, a hole is cut inside the charged plate so that some of the charged particles attracted to the plate pass through the hole. After passing through the hole, the charged particles keep going – with a natural tendency, i.e. their inertia, to continue traveling with constant velocity.

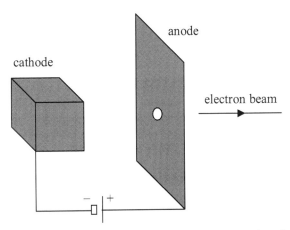

electrons jump from the negatively charged, heated cathode, accelerating through a potential difference supplied by a battery toward a positively charged anode (a collection plate) with a hole in its center; those electrons that pass through the hole form an electron beam

The charged particles passing through the hole form a beam of charged particles. By virtue of their inertia, no force is needed to keep them moving – they do that themselves. It may be convenient, though, to "store" the charged particles until they are needed. It is easy to store charged particles in a magnetic field because a charged particle in the presence of a uniform magnetic field either travels in a straight line, a helix, or a circle; magnetic forces only affect the direction of charged particles, and not how fast they move (unlike electric forces, which affect both). Thus, magnets (or current-carrying conductors) can be utilized to create magnetic fields that cause the charged particles to travel in circles – called storage rings. In the case of circular paths, the magnetic force $F_m = qvB$ supplies the needed centripetal force:

$$qvB = \frac{mv^2}{R}$$
$$qBR = mv$$

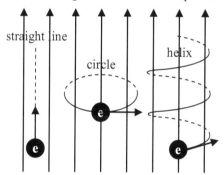

uniform magnetic field directed upward

straight line

circle

helix

an electron initially moving parallel or antiparallel to the magnetic field travels in a straight line (is unaffected by the magnetic field), since in this case the cross product of its velocity and the magnetic field is zero; an electron moving perpendicular to the magnetic field travels in a circle; otherwise, the path is a combination of the line and circle (a helix); the direction of the magnetic force is determined by the right-hand rule (and also accounting for the electron's negative charge)

After creating beams of charged particles, it is desirable to accelerate the particles to very high speeds, as this will result in higher-energy collisions. The charged particles can be accelerated to higher speeds by passing them through electric fields. When a power supply is connected across two conducting plates, the plates not only become charged, but also produce an electric field between the plates. The charged particles can be directed through such electric fields in order to increase their speeds. As the particles acquire higher speeds, they cover greater distances in shorter times, so magnetic fields are also needed to keep them contained. Here, the electric force $F_e = qE$ supplies the acceleration according to Newton's second law: $qE = ma$.

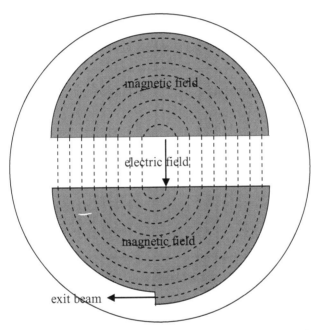

charged particles begin at the center of the cyclotron: an electric field accelerates charged particles in the central white region, while magnetic fields cause the charged particles to travel in circles in the shaded regions; each time the charged particle passes through the central white region, its velocity increases, and it travels in a wider circle in the shaded region until exiting with very high velocity

When the charged particles are originally captured to form the beam, they travel in small circles in uniform magnetic fields. At some point in their orbit, an electric field acts on them over a short time, increasing their speed and the radius of their orbit a little. This is repeated at some point in their second orbit, third orbit, and so on, until they have very high speeds and travel in large circles. As electrons have very little mass (inertia), they are very easily accelerated to very high speeds – nearly equal to the speed of light. They can travel $0.99c$ (or faster) in circles with radii of a mile or more.

When one clockwise and one counterclockwise storage ring intersect, the result is a number of high-energy collisions between charged particles. A group of detectors in the collision area look for the products of these high-energy collisions. The detectors are designed to determine which particles are produced in each collision and to allow determination of the velocities of the produced particles. As with production, it is easiest to detect charged particles. High-energy charged particles ionize atoms when they pass through matter in the detector, which leave distinct tracks that mark the particles' paths through the detector. The direction the particle curves indicates if it is positively or negatively charged, and precise measurements of its path reveal its speed. Photons produced during the collisions are also easy to detect through their interactions with atoms, and measurement of a photon's frequency provides its energy. Other neutral particles are more difficult to detect, but may be inferred directly – especially by measuring their decay products if the neutral particles decay within the detectors. For more details on the production and detection of elementary particles, see Ref. [A2].

The Large Electron-Positron (LEP) collider at the European Organization for Nuclear Research (CERN, from the French Conseil Européen pour la Recherche Nucléaire) in Geneva, Switzerland produced high-energy collisions between electrons and positrons. Although electrons and positrons commonly annihilate when they interact to produce two photons, there are a large number of possible outcomes from such a collision. It is also common for them to produce a new electron-positron pair. Very often they interact via the electromagnetic interaction with a photon as a mediator, but they can also interact via the weak interaction by exchanging $W^{\pm}$ or Z bosons. A large variety of final-state particles are possible when an electron and positron collide, such as a pair of charged leptons, a pair of neutrinos, a quark-antiquark pair, a pair of weak gauge bosons, and even three or more final states resulting from the same collision.

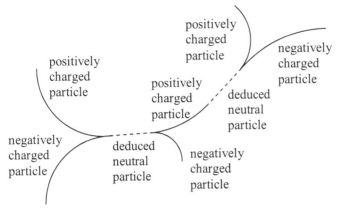

a simplified collision: negative and positive charges collide on the left; all charged particles leave tracks in the detectors, in which positive and negative charges travel in oppositely-directed circles and measurement of the radius of curvature determines the charge-to-mass ratio; connecting neutral particles must be deduced, as they are generally not observed directly

Like the rolling of dice at a craps table, the outcome of a single collision is uncertain, but if the experiment is repeated a large number of times, the observed outcomes fill out a probability distribution with much less uncertainty. The more the experiment is repeated, the less uncertainty there is in the observed outcomes. A standard die has 6 possible outcomes with equal probabilities – a $1/6^{th}$ chance of any side landing on top. This does not mean that a 1 will come out on top every 6 times a die is rolled. If a die is rolled 6 times, the outcome could very well be: 2, 3, 4, 3, 4, 6. Try it enough times and this result will surely happen at least once; "enough" could turn out to be a rather large number, though, but it may turn out to be small. If a die is rolled a small number of times, there is a fairly large chance that outcome will not be too evenly distributed. However, if the die is rolled a large number of times, there is a much higher probability that the distribution will be much more even. For example, roll a single die 120 times. Each side will probably not come out on top exactly 20 times each, but the outcome will likely be much more even than if it is rolled only 6 times. In the example above, the 3 and 4 came up twice, while the 1 and 5 did not show at all; that is a large discrepancy. Suppose that the die is rolled 120 times and the outcomes are: 18 (1), 21 (2), 24 (3), 21 (4), 17 (5), 19 (6). The 3 came up 7 more times than the 5, but no single face deviated from its expected value (20) by more than 20%. Compare to the previous example, where two-thirds of the faces deviated from their expected values (1) by 100%.

Collisions between particles do not produce even probability distributions, though – i.e. some outcomes have a much greater chance of occurring than others. Let us consider the case of rolling two die simultaneously as an example of producing an uneven probability distribution. Tossing two die at once, the possible outcomes are: both are 1, one 1 and one 2, one 1 and one 3, one 1 and one 4, one 1 and one 5, one 1 and one 6, one 2 and one 1, one 2 and one 2, etc. There are 36 possible pairings, counting a 2 and a 3 as different from a 3 and a 2, for example. Suppose that we are interested in the sum of the dice; then there are 11 possible outcomes – the dice can total as little as 2 and as much as 12. The dice are most likely to total 7, as there are 6 ways the dice can pair and add up to 7: 1, 6; 2, 5; 3, 4; 4, 3; 5, 2; 6, 1. That is, out of 36 possible pairings, $1/6^{th}$ add up to 7, which means that the probability of rolling two dice that add up to 7 is one-sixth. The least likely outcomes are 2 and 12, since there is only one way the dice can pair to total 2 and similarly for 12. Thus, the probability of rolling snake-eyes is $1/36^{th}$. This doesn't mean that if you roll two dice simultaneously 36 times that 6 times they will add up to 7 and one time they will add up to 12; rather, these are the 'expected values.' The outcomes will agree well with the expected values if the experiment is repeated a very large number of times, but may differ significantly if it is only repeated a small number of times. The probability distribution for tossing two dice simultaneously is illustrated below.

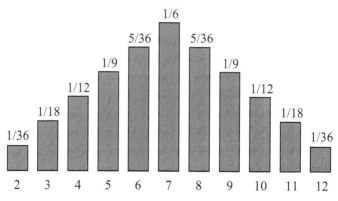

probabilities for possible outcomes of totaling two dice rolled simultaneously (assuming *fair* dice are used)

In the case of collisions between particles, the probabilities are ultimately determined by quantum mechanics (for which the main principles are introduced in Chapter 10). Feynman rules provide a convenient visual technique for drawing the possible outcomes, with a sophisticated mathematical process for calculating the probabilities. Theoretical particle physicists apply these Feynman rules to compute the probability distributions, while experimental particle physicists collide the particles repeatedly a large number of times, often for months or more, and measure the outcomes. The experimental observations are then compared with the theoretical predictions to check the level of agreement. While the mathematics associated with the Feynman rules can be quite complicated, there are a few general rules of thumb that can be useful to understand conceptually:

- An outcome that occurs at 'tree-level' is generally much more likely than an outcome that occurs at the 'loop-level.' The tree-level diagrams are simpler, and look like the branches of a tree; Feynman diagrams that include loops involve more propagators (particles that propagate a specific sub-interaction, called a vertex, of the collision).
- An outcome that is mediated by the strong interaction tends to be more likely than an outcome that is mediated by the electromagnetic or weak interactions, unless there are other compensating factors (like loops in one diagram, but not the other).
- An outcome that features more vertices tends to be less likely than an outcome that is represented by a Feynman diagram with fewer vertices, all else being equivalent.
- An outcome that results in more massive particles is generally less likely than an outcome that results in lighter final-state particles – depending, of course, also on the number of loops and vertices and types of interactions involved.

So, for example, if electrons and positrons are smashed together 100,000 times over the course of a few months, it is possible to calculate how many times we expect to see an outcome of two photons, an outcome of an electron and positron, an outcome of a quark-antiquark pair, etc. Suppose that theory predicts that a particular outcome will occur 4.3% of the time. If the measurement reveals that the outcome instead occurs 3.9% of the time, does this mean that the theory is incorrect? This question is very fundamental to analyzing the results. As illustrated with the dice, we don't expect to get *exactly* the expected number of outcomes, but for a large number of experiments, we expect to get *close* to the expected value. But what is close enough? Statistically, there is a good prescription for quantifying the level of agreement: Computing the standard deviation is one way to determine the uncertainty – i.e. how far we can expect to be off.

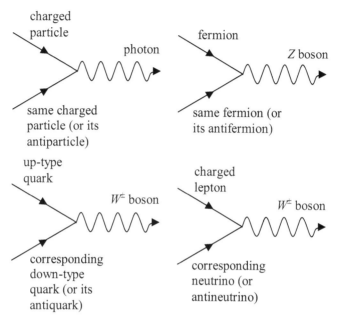

fundamental Feynman vertices for electroweak interactions

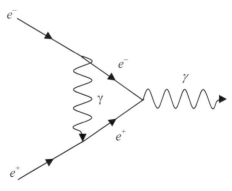

a loop-level Feynman diagram

Incorporating the statistics into the analysis, the theoretical prediction would be of the form 4.3 ± 0.3%. The 0.3% is the standard deviation. If the results agree within one standard deviation – i.e. if the experimental value lies between 4.0% and 4.6%, there is excellent agreement; if the results agree within three standard deviations – in this case, between 3.4% and 5.2%, there is still agreement, but not as good. If the experimental value lies outside of the three-sigma (i.e. three standard deviations) range, there is a clear discrepancy. The closer the agreement, the higher our confidence level – or, the greater the disagreement, the greater our confidence level that they do not agree. The confidence level can also be quantified statistically. Much of the analysis is based on 90% confidence levels.

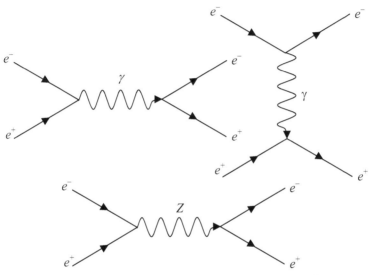

some Feynman diagrams where an electron and positron interact and produce an electron and positron in the final state

Considering all of the uncertainties involved, difficult computations, experimental challenges, the probabilities underlying the quantum mechanics, and so on, the level of agreement between the Standard Model's theoretical predictions and the experimental outcomes for the LEP collider is quite amazing [T6]: A large number of the measurements agree at the one-sigma level, within relatively small standard deviations, and the vast majority agree within the three-sigma level. The outliers are too few to offer more than speculation – i.e. there is no conclusive evidence for new physics (beyond the Standard Model).

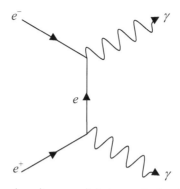

an electron and positron annihilate, producing two photons

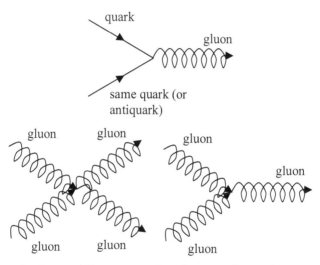

fundamental Feynman vertices for strong interactions

With a maximum collider energy of 209 GeV (10^9 electron-volts, where an electron volt is the energy needed to move an electron through a potential difference of one volt), the LEP collider had the highest beam energy for colliding leptons. It began operation in 1989 and concluded in 2000. The electron and positron beams traveled through circular underground tunnels on the border between France and Switzerland, which had a diameter of about 9 kilometers. The LEP successfully discovered the $W^{\pm}$ and Z bosons predicted by the electroweak theory that features the Higgs mechanism, which is why there is now large anticipation for the discovery of the Higgs.

The Tevatron is the most powerful collider in the United States, located at the Fermi National Accelerator Laboratory in Illinois. The Tevatron produces collisions between protons and antiprotons with a collider energy of up to 1.96 TeV (10^{12} eV) with rings over a mile in diameter. As protons and antiprotons consist of quarks and antiquarks, the majority of their collisions involve the strong nuclear force, whereas the collisions between electrons and positrons at the LEP collider chiefly involved the electromagnetic or weak interactions. This results in a much different set of probable outcomes for collisions at the Tevatron. Although protons collide with antiprotons at the Tevatron, the underlying interactions are generally between one quark from the proton and one antiquark from the antiproton, which involves modeling the proton and antiproton as composite particles – a theoretical complication that can be modeled fairly well, but adds a little uncertainty to the predicted probability distributions. The Tevatron successfully discovered the missing third-generation particles – namely, the top and bottom quark and the tau lepton neutrino.

Returning to the LEP collider, it was shut down in 2000 to make room for the Large Hadron Collider (LHC). The LHC is presently the world's highest-energy collider, colliding protons together at energies of up to 14 TeV. The proton beams were successfully run in the summer of 2008, but then halted to repair faults in the superconducting magnets; the LHC is expected to be operational in the summer of 2009. The main expectations for the LHC include:

- discovery of the Higgs boson(s) and determination of the mass(es) of the Higgs boson(s), or the surprising discovery that the Higgs does not exist (at least not as presently theorized)
- testing the theory of supersymmetry to see if there are very massive supersymmetric counterparts for the observed Standard Model particles in our universe (see Chapter 10)
- developing a better understanding of the hierarchy problem (discussed in Chapter 8), and whether or not there are actually two fundamentally different scales in nature
- discovering large extra dimensions associated with superstring theory (and at the same time provide evidence for superstring theory), or placing bounds on how large such extra dimensions can be

The LHC is expected to discover one or more new particles; it will be a major surprise if the LHC does not, and will send all of the theorists back to the drawing board. Discovering a new particle at a collider is not an easy task. When a stable, charged particle like the electron is produced, it leaves a very clear signature – i.e. the curved path it takes through the detector can be analyzed to determine that the particle has the properties of an electron. However, most particles, especially those that the LHC is expected to discover, are not stable – i.e. they decay to lighter particles in a very short time period, usually within the detector.

For example, when the Tevatron produced top quarks in proton-antiproton collisions, the top quarks quickly decayed to lighter particles – one of the more probable outcomes being a bottom quark and a W boson (as with collisions, the probability distributions for decays are governed by the Feynman rules). To discover the top quarks, it was necessary to analyze the tracks left in the detector after each collision to look for groups of particles that would be left after the top quarks decayed – such as looking for bottom quarks and W bosons (which themselves decay to other particles). Such reconstruction is a tedious process (for a large number of similar events), both theoretically and experimentally, but teams of a hundred or more particle physicists typically collaborate on such tasks and it is reasonably straightforward to work out both the theoretical and experimental probability distributions for comparison. There is also a matter of determining the backgrounds – i.e. other processes that lead to the same final states without producing top quarks – and subtracting the backgrounds from the observed number of events. The top quark and other particles have been discovered through this somewhat indirect method, with very high confidence levels – i.e. there is no doubt that the top quark has been discovered, as it was produced in abundance with all the properties that the top quark is expected to have.

If the LHC discovers new particles, it will most likely be through a similar process. If new particles are produced in abundance, they will be very highly accepted and understood. It is also possible that a new particle, or what seems like a new particle, may be produced only a few times, in which case there will be more uncertainty about whether or not it is a new particle or if there are a few more backgrounds than were theoretically anticipated. So, for new particles to be truly discovered, more than just a few may need to be produced. This is why colliders may run for several years – to repeat the collision so many times that the uncertainties will be minimal.

Lighter particles are more likely to be produced in abundance; heavier particles are more difficult to produce, generally. So, if a new particle is not produced at a collider, it does not necessarily mean that it does not exist: It could mean that the particle is very heavy. This is why the LEP collider was unable to discover the Higgs boson or top quark, why the higher-energy Tevatron was able to discover the top quark, and why we expect the yet higher-energy LHC to have enough energy to discover the Higgs boson and other possible particles. Some particles, like the Higgs boson, have theoretical upper limits on their masses, meaning that if the LHC does not discover the Higgs, theorists will need to develop a new mechanism for giving masses to the Standard Model particles. Other particles, like Kaluza-Klein excitations, for example, may be too heavy to be discovered at the LHC, but also may be light enough to be discovered. In this case, the particles will either be discovered by the LHC or their lack of discovery will raise the lower bounds on their possible masses.

Experimental evidence for mechanisms of new physics can also come without the direct discovery of a new particle: New particles or interactions can affect the production rates for ordinary particles. So when the theoretical probability distributions, based on the Standard Model, are compared with the experimental outcomes, a notable discrepancy, such as three-sigma or more, may be a sign of new physics. In this case, theorists will apply models beyond the Standard Model, such as Grand Unified Theories, to see if they improve agreement between theory and experiment. Although the LEP collider and Tevatron have not left much room for new physics, at their lower energies, new physics mechanisms (such as supersymmetry) would not have much effect, whereas they are expected to have a more discernible effect at the larger energies of the LHC.

We can actually improve over the comparison of probability distributions to probable outcomes that we have discussed because the experimentalists do more than tabulate sets of final states for each collision: They actually make high-precision measurements of the momentum and energy of the final-state particles (that of the initial particles being already known based on the beam properties). Many of the sources of possible new physics that the LHC will be looking for involves very heavy particles (otherwise, if they were lighter, in most cases we would expect to have observed them at the LEP collider or Tevatron), and heavy particles have a high probability of leading to final states (through subsequent decays within the detector) with high amounts of transverse momentum and energy. As a vector, momentum has both direction and magnitude: It has a longitudinal component along the beam axis and a transverse component perpendicular to the beam axis. Compared to the production of lighter particles, heavier particles tend to have more transverse momentum. This type of filter is called a transverse momentum cut. Finding a significant number of events with final states that have high amounts of transverse momentum can be a signal for a source of new physics; in this case, the

Standard Model backgrounds are usually relatively small, since the Standard Model particles are rather light (except for the top quark, which can be identified since it is very well understood now) compared to new particles that the LHC will be looking for – i.e. the Standard Model processes usually lead to final states with comparatively low amounts of transverse momentum.

Thus, the LHC will look for sources of new physics by (1) looking for new particles, probably by reconstructions from tracks left by final-state particles left after heavy new particles decay in the detector; (2) closely comparing the observed outcomes of collisions to theoretical probability distributions for possible evidence of new physics; and (3) filtering through the data with criteria such as large amounts of transverse momentum, which would tend to make a signal for new physics stand out against the Standard Model backgrounds. If the LHC detects the presence of large extra dimensions in our universe, it will probably be through one of these methods. If the LHC does not discover extra dimensions, then these three methods will be employed to set bounds on how large any extra dimensions can be.

9.3 Collider Searches for Extra Dimensions

If there are extra dimensions in our universe, any particle that propagates into one or more extra dimensions will have associated Kaluza-Klein excitations (as described in Chapter 8). The LHC and other upcoming high-energy colliders will look for evidence of such Kaluza-Klein excitations. Data from other current high-energy colliders, like the Tevatron, are also being analyzed for evidence of large extra dimensions, as well as data from prior colliders, like LEP. Since the LHC will be running at much higher energies, it has better chances of discovering extra dimensions, so we will concentrate our attention on the LHC. However, many of the collider searches that we will describe for collisions between protons at the LHC are also applicable to other colliders, including any higher-energy colliders that may come in the future.

As with other sources of new physics, the LHC can search for the direct production of Kaluza-Klein excitations, which really means looking for final-state particles that may have decayed from Kaluza-Klein excitations, or for indirect effects that Kaluza-Klein excitations would have on the probability distributions, which can be amplified by filtering for such criteria as high amounts of transverse momentum in the final-state particles. As mentioned in Chapter 8, the theoretical predictions are somewhat model dependent. Three main models include: (1) Models in which only gravitons propagate into extra dimensions, (2) models in which gauge bosons can propagate into one or more extra dimensions (in addition to gravitons), and (3) the UED model where all of the particles can propagate into one or more extra dimensions. Further refinements include the number and shape of the extra dimensions.

Let us first consider the case where only gravitons propagate into extra dimensions, while the Standard Model particles are confined to the usual D_3-brane. In this case, only gravitons have Kaluza-Klein excitations. A single event involving a graviton is extremely weak compared to other events because gravity is by far the weakest of the fundamental forces (at least, at energy scales that have been probed experimentally – at some higher energy scale, the four forces may turn out to be different manifestations of a single underlying force). Two protons interacting with one another electromagnetically experience a repulsive force that is about 10^{36} times stronger than their gravitational attraction.[8] When two high-energy protons collide, final-state outcomes that arise due to gravitational interactions are generally vastly insignificant compared to final-state outcomes arising from strong, electromagnetic, or weak interactions; the gravitational interactions generally have no discernible effect on the probability distributions. At least, that is the case in the Standard Model and most sources of new physics, but large extra dimensions provide an exception to this rule.

Ordinarily, two high-energy protons can collide by interacting with a gluon, photon, $W^{\pm}$ boson, Z boson, or graviton, in which the interaction involving the graviton pales in comparison with the other forces. However, in the case of large extra dimensions, there is not only the graviton, but an associated tower of Kaluza-Klein excitations of the graviton. This tower of Kaluza-Klein excitation can be, in the case of the graviton, a very large number of particles that behave just like the graviton except for having rest-mass. In this case, the colliding protons can interact by exchanging a graviton or any of its numerous Kaluza-Klein excitations, and the very large number of possible mediators of the gravitational interaction can amplify the signal for the gravitational interaction significantly – enough to observe its effect at a high-energy collider.

---

[8] This factor is different for electrons, which have less mass than protons.

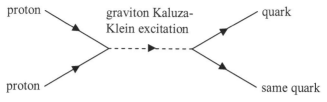
two protons collide and produce a pair of quarks through the exchange of the Kaluza-Klein excitation of a graviton

Computationally, however, it is not as simple as calculating the contribution for a single graviton exchange and multiplying by the number of Kaluza-Klein excitations of the graviton in the tower. The reason is that the Kaluza-Klein excitations do not all provide equal contributions because they differ in mass. The graviton is massless, whereas its Kaluza-Klein excitations effectively have tree-level[9] rest-masses $m_1 = m_K$, $m_2 = 2m_K$, $m_3 = 3m_K$, and so on (a special case of the formula given in Chapter 8, in which the Standard Model particle is massless), where m_K is inversely proportional to the compactification radius. The graviton exchange gives the greatest contribution, since it is massless, its first Kaluza-Klein excitation contributes slightly less, then its second Kaluza-Klein excitation, and so on, with its highest-level Kaluza-Klein excitation contributing much, much less because it has so much more mass. Nonetheless, these contributions can be added together very efficiently, even if there is an extremely large number of Kaluza-Klein excitations in the tower, since there are straightforward mathematical techniques for adding numbers that follow a simple series. For example, the numbers 1 thru 1,000,000 can be added together in very little time – without a calculator! – by realizing that 1 + 999,999 equals 1,000,000, that similarly 2 + 999,998 equals 1,000,000, and so on such that there are 499,999 pairs of numbers in the series that add up to 1,000,000 plus the unused numbers 500,000 and 1,000,000: So there are 500,000 millions plus 500,000 which totals 500,000,500,000.

Puzzle 9.1: Add each of the following without a calculator:
- the numbers 1 thru 20
- the numbers 100 thru 200

---

[9] The Kaluza-Klein excitations also receive mass contributions corresponding to loop-level Feynman diagrams. As a general rule, the tree-level contributions tend to provide the main contribution.

- the even numbers from 2 to 200

Direct tests of Newton's law of gravitation show that extra dimensions accessible only to gravitons may be as large as about a tenth of a millimeter, which corresponds to $m_K \approx 10^{-3} \text{ eV}/c^2$. (In particle physics, it is convenient to express masses in terms of energy, in electron·volts, and the speed of light squared, for easy comparison between mass and energy from Einstein's famous equation, $E = mc^2$. In this way, it is easily seen that a mass of $10^{-3} \text{ eV}/c^2$ corresponds to an energy of 10^{-3} eV, where one electron·volt is the energy required to move a proton through a potential difference of one volt; while for an electron the energy would be the same, but negative.) If the extra dimensions are as large as about a tenth of a millimeter, then the first Kaluza-Klein excitation of the graviton would have a mass of about $10^{-3} \text{ eV}/c^2$, the second Kaluza-Klein excitation of the graviton would have double this mass, the third Kaluza-Klein excitation of the graviton would have triple the mass of $10^{-3} \text{ eV}/c^2$, etc. Based on tests of Newton's law of gravitation, no Kaluza-Klein excitation can be lighter than $10^{-3} \text{ eV}/c^2$; however, the lowest-lying Kaluza-Klein excitation of the graviton may be much heavier than this. The smaller m_K, and hence the larger the compactification radius, the greater the phenomenological effects of the extra dimensions, so the lower limit on m_K and the upper limit on the compactification radius are important bounds to work with as they yield the best chances of detecting the presence of large extra dimensions.

Recall, from Chapter 8, that the motivation for large extra dimensions – i.e. much larger than the Planck length – stems from the concept of a single fundamental scale in nature, on the order of the observed electroweak scale, which is about 10^{12} eV. In this case, the energy scale 10^{12} eV also serves as the fundamental scale for gravity, or the effective string scale (instead of the traditional Planck scale). As a result, this fundamental scale 10^{12} eV provides an approximate upper limit for the masses of Kaluza-Klein excitations (technically, it may extend past this somewhat by about a factor of up to 10 or 100). Thus, the heaviest Kaluza-Klein excitation of the graviton must have a mass of about $10^{12} \text{ eV}/c^2$.

If the lowest-lying Kaluza-Klein excitation of the graviton has a mass of $10^{-3} \text{ eV}/c^2$, there are about 10^{15}, or 1,000,000,000,000,000 Kaluza-Klein excitations of the graviton (actually, as we will explore shortly, this number is actually much higher as a result of degeneracy, but this is a useful starting point). While a single collider event involving a graviton makes a miniscule contribution to the probability densities, which is usually extremely insignificant, the presence of 10^{15} Kaluza-Klein excitations of the graviton that result from the presence of large extra dimensions can yield a substantial contribution. If the compactification radius is smaller, such that m_K is larger than 10^{-3} eV$/c^2$, then there are fewer Kaluza-Klein excitations in the tower and the effect of the extra dimensions is somewhat smaller; but it can still be quite significant unless the radius is considerably larger.

In the case of two or more extra dimensions of the same size accessible only to gravitons, there is a degeneracy of Kaluza-Klein excitations. With a single extra dimension, the Kaluza-Klein excitations are numbered $n=1$, $n=2$, $n=3$, and so on. For two extra dimensions, two indices are needed to number the Kaluza-Klein excitations, which is related to the fact that the momentum of a graviton propagating into two extra dimensions has two independent components of momentum in the extra dimensions – or two hidden degrees of freedom. In this case, the Kaluza-Klein excitations correspond to ordered pairs (0,1), (1,0), (1,1), (0, 2), (2,0), (1, 2), (2,1), (2,2), (0,3), (3,0), etc., and have tree-level masses $m_{0,1} = m_{1,0} = m_K$, $m_{1,1} = \sqrt{2} m_K$, $m_{0,2} = m_{2,0} = 2 m_K$, $m_{1,2} = m_{2,1} = \sqrt{5} m_K$, $m_{2,2} = 2\sqrt{2} m_K$, $m_{0,3} = m_{3,0} = 3 m_K$, etc., according to the formula $m_{n_1,n_2} = \sqrt{n_1^2 + n_2^2} \, m_K$. The number of states grows quickly with an increasing number of extra dimensions, with a growing degeneracy of Kaluza-Klein masses – i.e. more than one combination of indices corresponding to the same Kaluza-Klein mass. In fact, the number of Kaluza-Klein excitations grows so rapidly as the number of extra dimensions increases that this number is very sensitive to the cut-off scale, which sets the upper limit for Kaluza-Klein masses. While the observed electroweak scale is on the order of 10^{12} eV, the cut-off scale may be as high as about 10^{14} eV.

Therefore, the effect that the presence of large extra dimensions accessible only to gravitons has on the probability distributions for collider physics is sensitive to three main parameters: the size of the extra dimensions, the number of extra dimensions, and the ratio of the cut-off scale to the actual electroweak scale. Superstring theory predicts six extra dimensions (and effectively a seventh in M-theory), but the extra dimensions may be asymmetric – i.e. they might not all be of the same size, so there could be a couple of large extra dimensions and a few that are much smaller. Evidently, all of the dimensions are not the same size – three are macroscopic and the others are apparently compact – so why should all of the extra dimensions have the same size? On the other hand, why shouldn't they be? Perhaps this is related to other issues, like why first-, second-, and third-generation quarks have much different masses – i.e. theoretically, it may be possible to link asymmetric extra dimensions to asymmetric masses. This doesn't mean that extra dimensions should be asymmetric – just that they could be. It is important to keep in mind that different scenarios are possible.

Examination of current and prior collider data leads to constraints on these parameters: how large extra dimensions can be, depending upon how many large extra dimensions there are and the ratio of the cutoff scale to the electroweak scale. A plot of these parameters illustrates the parameter space, and collider and other data show which regions of the plot are allowed or excluded. Data from current and prior colliders affect the allowed and excluded regions, but do not yet strengthen the bound on the size of extra dimensions over direct tests of Newton's law of gravitation. So the bound for large extra dimensions accessible only to gravitons remains at about 0.1 mm [T2-T3]; this limit is on the verge of probing both by high-energy colliders and by direct tests of Newton's law of gravitation.

Observe that the electroweak scale is on the order of 10^{12} eV, or a TeV for short. Prior colliders probed energies in the hundred GeV range. For example, the LEP collider ran at 200 GeV, corresponding to 0.2 TeV. These energies were too low to have good prospects for discovering mechanisms of new physics occurring at the TeV scale. The Tevatron is running at 2 TeV, which is right on the cusp of possibly discovering new physics at the electroweak scale. As a result, the Tevatron data is being analyzed, as it is being gathered, for possible signs of new physics at the electroweak scale, including evidence for the presence of large extra dimensions in our universe. The Tevatron has a chance for observing this, but its energy may turn out to fall a little short of what is needed. Soon to be up-and-running at energies of up to 14 TeV, all eyes are on the LHC for good prospects of discovering new physics in the near future; and if this falls a little short, the coming generation of colliders can be expected to probe somewhat further. The current and upcoming colliders, which may probe energies up to about 100 TeV, offer excellent prospects for developing a full understanding of electroweak physics, including any relation that it may have to an effective string scale in the 1-100 TeV range.

Thus far, we have considered collider searches for extra dimensions that are accessible only to gravitons. Now we will look at the other side of the spectrum and consider collider searches for extra dimensions in the UED model. Both may very well be relevant: In the context of asymmetric extra dimensions, there may be one or more dimensions as large as a sub-millimeter into which only gravity can propagate in addition to one or more extra dimensions into which all of the Standard Model particles can propagate. However, extra dimensions accessible to ordinary particles cannot be as large as 0.1 mm; as we will see, they must be much smaller.

In the UED model, all of the Standard Model particles can propagate into one or more extra dimensions, but not necessarily all of the extra dimensions: The Standard Model particles reside on a $D_{3+\delta}$-brane, where δ is the number of extra dimensions into which the Standard Model particles have access. For now, we will consider $\delta = 1$, for simplicity. All of the Standard Model particles have associated Kaluza-Klein excitations with tree-level masses $m_1 = \sqrt{m_S^2 + m_K^2}$, $m_2 = \sqrt{m_S^2 + 4m_K^2}$, $m_3 = \sqrt{m_S^2 + 9m_K^2}$, and so on, where m_S is the rest-mass of the Standard Model particle, m_K is an effective mass that represents, naïvely, the hidden energy (from our macroscopic 3D perspective) from the Standard Model particle's motion in the extra dimension (which, as we have seen, is inversely proportional to the radius of the extra dimension), and m_1, m_2, m_3, and so on are the masses of the Kaluza-Klein excitations of the Standard Model particle.

When a single Standard Model particle travels through the higher-dimensional D_4-brane, a 4D space (really, a spacetime with four dimensions of space and one dimension of time), we would observe the particle differently from our macroscopic 3D perspective, since it has hidden energy associated with its component of momentum along the extra dimension. Mathematically, the single particle in the higher-dimensional spacetime would be treated as a Fourier series, which includes one term for a Standard Model particle (with its usual mass) traveling in the usual 3D space plus a tower of Kaluza-Klein excitations, which are effectively heavier particles. In a collider, we would see an effectively 3D interaction involving either 3D Standard Model particles or one of their associated Kaluza-Klein excitations. (Technically, we treat the interactions in a relativistic spacetime continuum, so one dimension of time must inherently be added to each of these interactions. Nonetheless, we will continue to focus on the spatial dimensions.)

In contrast to the case where only gravitons propagate into the extra dimensions, extra dimensions that Standard Model particles propagate into cannot have such a large size so as to correspond to a small value of m_K. Recall that m_K can be as large as 10^{-3} eV$/c^2$ in the case of extra dimensions accessible only to gravitons. This value of m_K is immediately ruled out for extra dimensions accessible to Standard Model particles, since it is not large compared to the masses of the Standard Model particles. Consider, for example, the electron, which has a mass of 0.511 MeV$/c^2$. If the electron were to propagate into an extra dimension with a radius large enough that m_K is not very large compared to the electron's mass – such as 10^{-3} eV$/c^2$, but even up to several MeV$/c^2$ – then there would be a large tower of Kaluza-Klein excitations that are exact replicas of the electron except that, in this case, the lowest-lying Kaluza-Klein excitations would be very light. The existence of such light Kaluza-Klein excitations of the electron would have already been detected at the LEP, Fermilab, and other colliders. It is possible for Kaluza-Klein excitations of the graviton to be as light as 10^{-3} eV$/c^2$ because the gravitational interaction is very weak compared to the strong, weak, and electromagnetic interactions, but Kaluza-Klein excitations of the Standard Model particles – especially leptons, quarks, and photons – must have a mass of at least about 350 GeV$/c^2$ in order to explain why their associated Kaluza-Klein excitations have not yet been detected with existing collider data. Since 100 GeV$/c^2$ equates to 10^{11} eV$/c^2$, the bound on the size of extra dimensions into which all of the Standard Model fields can propagate is about 10^{11} times stronger (about 10^{-15} m) than the bound for extra dimensions accessible exclusively to gravity (about 0.1 mm). This means that electrons, photons, and other ordinary particles cannot propagate more than about the size of an atomic nucleus from the ordinary dimensions, which easily explains why we are unable to move into extra dimensions or even "see" particles moving in that direction. Because the extra dimensions must be so small if ordinary particles can propagate into them, we need experiments between microscopic particles occurring at very high energies, such as through collisions at high-energy colliders, in order to have a plausible chance of detecting them.

In the UED model, the probability distributions for possible final states resulting from high-energy collisions are affected by the Kaluza-Klein excitations of the usual SM particles. In particular, when two ordinary particles are smashed together in a high-energy collider, there will be a probability of producing two or more Kaluza-Klein excitations. It is not possible to produce just one Kaluza-Klein excitation of the SM particles in the UED model as a result of conservation of 4D momentum. When two ordinary particles collide, neither particle has momentum along the extra dimension; whereas a Kaluza-Klein excitation produced as a result of the collision carries a component of momentum in the hidden dimension. Since momentum is conserved for collisions between particles, the total momentum of the final-state particles can not have a component along the extra dimension. Therefore, Kaluza-Klein excitations must be produced in pairs (or in groups of three or more), so that their components of momentum along the extra dimension cancel out.

The collider signature of the UED model is the production of two Kaluza-Klein excitations resulting from the high-energy collision of two beam particles (e.g. two protons for the LHC or a proton and antiproton at Fermilab). This final state needs to occur with a probability that is not so low as to be statistically insignificant. (If the experiment continues for years and only two such events are observed, the result could be a statistical outlier.) There is not just a matter of producing them, but identifying the Kaluza-Klein excitations in the detector. Just exactly what is observed in the detector depends on the properties of the particles produced and whether or not they decay within the detector. Identification is more difficult when the particles decay, as there may be other particles that result in similar final states after decaying. So there is not only the matter of producing the Kaluza-Klein excitations, but also detecting them.

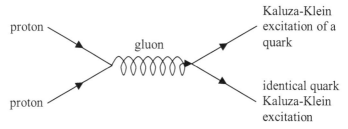

two protons collide and produce a pair of Kaluza-Klein excitations through the exchange of a gluon

The difficulty in producing a pair of Kaluza-Klein excitations in the UED model is that they are inherently very heavy. As mentioned earlier, Kaluza-Klein excitations must have a mass of at least 350 GeV$/c^2$ – otherwise, they would have been detected by now. (Recall that the energy of the particle, in giga electron-volts, or billions of electron-volts, is related to its mass through Einstein's equation, $E = mc^2$, where c is the speed of light.) Conservation of energy requires a minimum of 700 GeV of energy from the colliding particles in order to produce a pair of Kaluza-Klein excitations in the UED model. Higher-energy colliders are required to produce heavier particles; plus, heavier particles tend to decay rapidly, making them difficult not only to produce, but also to detect. The LHC, with a collider energy of 14 TeV (which is 20 times 700 GeV), has plenty of energy to either discover Kaluza-Klein excitations in the UED model or significantly raise the current mass bound of 350 GeV/c^2 (to a few TeV).

Associated with conservation of momentum in 4D is what is referred to as conservation of Kaluza-Klein number – which corresponds to the component of momentum along the extra dimension. Kaluza-Klein number is conserved in the UED model. Feynman diagrams for possible collisions and decays involve initial- and final-state particles, plus virtual particles that mediate the interactions – called propagators. Particles usually interact in threes (one of the rare exceptions is four gluons interacting together). Each particle has a Kaluza-Klein number – zero for a Standard Model particle, one for the first excitation, two for the second excitation, and so on. For an interaction between three particles (a vertex in a Feynman diagram), Kaluza-Klein number conservation requires that $m \pm n = \pm k$, where these are the Kaluza-Klein numbers of each particle meeting at the vertex.

Suppose that the particle corresponding to m is not a Kaluza-Klein excitation. In this case, $m = 0$, which requires that $n = k$. This means that if one particle is a Standard Model particle, the other two particles must have the same Kaluza-Klein number. We can also see that there must either be zero, two, or three Kaluza-Klein excitations meeting at a vertex: If there are two Standard Model particles, corresponding to m and n, for example, then k must also be zero (since $0 \pm 0 = 0$). Tree-level conservation of Kaluza-Klein number is the reason that Kaluza-Klein excitations must be pair-produced (and so can not be produced singly).

Puzzle 9.2: Which of the following interactions (vertices) conserve Kaluza-Klein number?

$$u_1^* \bar{u}_4^* g_5^* \qquad d\bar{d}_1^* g_2^* \qquad g_2^* g_5^* g_3^*$$

$$t_6^* \bar{t}_{16}^* g_1^* \qquad g_2^* g_1^* g \qquad g_2^* g_6^* g_5^* g_1^*$$

Another consequence of Kaluza-Klein number conservation is that Kaluza-Klein excitations are stable at the tree-level. For example, consider the first Kaluza-Klein excitation of the b quark, the b_1^*. The b_1^* can couple to a Standard Model b quark, designated simply as b, and the first Kaluza-Klein excitation of the gluon, the g_1^*. The gluon is massless, so the g_1^* has a mass of m_{KK}, which is the mass associated with the size of the extra dimension (some value greater than about 350 GeV$/c^2$). The b has a mass of $m_b = 4.2 \text{ GeV}/c^2$. Applying a formula that was stated earlier, the b_1^* has a mass of $m_{b_1^*} = \sqrt{m_b^2 + m_{KK}^2}$. You should recognize that this formula has the structure of the Pythagorean theorem. When applied to a right triangle, the Pythagorean theorem states that the hypotenuse, which has a length of $c = \sqrt{a^2 + b^2}$, has a length that is shorter than the sum of the two sides: $c < a + b$. Analogously, $m_{b_1^*} < m_b + m_{KK}$. As a result, any possible decay of the b_1^* to a g_1^* and a b is kinematically suppressed – i.e. the b_1^* can not decay this way because it does not have enough mass. Similarly, none of the Kaluza-Klein excitations can decay at tree-level without violating conservation of 4D momentum.

This means that one or more of the variety of the lowest-lying Kaluza-Klein excitations may actually be stable – they may not decay at all. However, loop-level calculations show that there can be mass splittings between the Kaluza-Klein excitations of a given level to allow for many of the Kaluza-Klein excitations to decay. They will decay to the lightest Kaluza-Klein particle. For example, if the lightest Kaluza-Klein excitation is the photon, the Kaluza-Klein excitations of charged particles will decay to Standard Model particles and Kaluza-Klein excitations of the photon. In this case, when Kaluza-Klein excitations are pair-produced, the detector will see two ordinary particles – which will look very much like a Standard Model process. This is not as obvious as observing new particles in the detector, but can still have a significant effect. If a significant number of Kaluza-Klein excitations are produced during high-energy collisions, this will result in an excess of various Standard Model final states. So if more events of various Standard Model final states are observed than theoretically expected, this can be the result of extra dimensions in the UED model. Also, final-state particles that are produced from decays of Kaluza-Klein excitations will have large amounts of transverse momentum, which can stand out relative to the Standard Model background (ordinary processes leading to the same final states). Furthermore, there will be missing energy from particles that are not observed in the detector. So, after taking time to analyze the individual signals for a large number of collisions, measurements of transverse momentum and missing energy can help to detect the possible presence of UED-type extra dimensions in our universe.

Puzzle 9.3: For each decay listed below, determine if it is strictly forbidden or if it may be possible in the case where there are mass splittings between the Kaluza-Klein excitations of a given level:

$$t_1^\bullet \to t + g \qquad t_1^\bullet \to b_1^\bullet + W^+ \qquad c_1^\bullet \to s_1^\bullet + W^+$$

$$g_1^* \to t + \bar{t} \qquad u_4^\bullet \to u_2^\bullet + g_1^* \qquad t_1^\bullet \to t + \gamma_1^*$$

As we saw with gravitons, the number of extra dimensions can have a large effect on the bounds for the size of the extra dimension (and hence the masses of the Kaluza-Klein excitations) because there is a growing degeneracy of Kaluza-Klein excitations with an increasing number of extra dimensions. That is, with two or more extra dimensions, there are more Kaluza-Klein excitations that would be produced during high-energy collisions. In the case of two or more extra dimensions that are universal, there is an additional effect that is rather curious: The spin angular momentum of particles may be very different in higher-dimensional space.

Puzzle 9.4: In the case of $\delta = 3$, how many distinct types of Kaluza-Klein excitations of the up-type quark, $u_{m,n}^\bullet$, are there with masses less than $5m_{KK}$? (Neglect any mass splittings.)

Let us recall what we have learned about spin. Particles can have two kinds of angular momentum – orbital angular momentum, analogous to the revolution of the earth around the sun, and intrinsic spin angular momentum, like the earth's rotation about its axis. Spin angular momentum is said to be an intrinsic property (at least, from the vantage point of 3D space, in the context of pointlike particles): A pointlike elementary particle cannot physically "spin." A moving charged particle creates a magnetic field, yet a stationary charged particle that has spin also creates a magnetic field by virtue of its spin angular momentum. With respect to spin, there are two kinds of particles – fermions and bosons. Fermions have half-integer spin

($\pm\hbar/2$, $\pm 3\hbar/2$, $\pm 5\hbar/2$, etc.), while bosons have integral spin (0, $\pm\hbar$, $\pm 2\hbar$, $\pm 3\hbar$, etc.). Common fermions include charged leptons (the electron, muon, and tau), neutrinos, quarks, and baryons (including the proton and neutron). Mediators like the photon, graviton, gluons, and $W^{\pm}$ and Z are bosons. Mesons, consisting of a quark-antiquark pair, are also bosons. Bosons with zero spin are scalars, while bosons with nonzero spin are vector particles.

In 3D, fermions are observed to behave as spinors with half-integral spin, which may be spin up or spin down. Fermions obey the Pauli exclusion principle – no two fermions can have the same set of quantum numbers (no two electrons in an atom can exist in the same state; if they reside in the same subshell, one will be spin up while the other will be spin down). As a consequence of the extra geometric degrees of freedom in higher-dimensional space, fermions would instead behave as vector-like particles if they propagate in a 4D space. In higher dimensions, fermions would exist as spinors if there are an odd number of dimensions (i.e. 5D, 7D, etc.) and as vector-like particles if the number of dimensions is even (4D, 6D, etc.).

In 3D, fermions are two-component Weyl spinors. They have left-chiral and right-chiral states. The left-chiral fermions couple to the $W^{\pm}$ bosons, but the right-chiral fermions do not; both left-chiral and right-chiral fields may couple to the other gauge bosons (respecting usual constraints – neutral fermions, for example, do not couple to the photon, and only quarks couple to gluons). Mathematically, the left-chiral fermions form the group $SU(2)$, while the right-chiral fermions form the group $U(1)$. Left-chiral fermions form doublets – e.g. the left-chiral electron and left-chiral electron neutrino together couple with the $W^{\pm}$, and the left-chiral up and down quarks similarly form a doublet through their interaction with the $W^{\pm}$ – whereas the right-chiral fermions form singlets.

Generalizing particle physics to higher-dimensional space, in even-dimensional spaces the fermions would be four-component vector-like particles, while in odd-dimensional spaces the fermions would be spinors. In 4D, 6D, and so on, both the left-chiral and right-chiral fermions couple to all of gauge bosons, including the $W^{\pm}$: There are both left-chiral and right-chiral doublets and singlets in this case.

If fermions propagate into an even-dimensional space in our universe – as would be the case in the UED model with one, three, or five extra UED-type dimensions – in the effective 3D theory, the fermions must still behave as two-component Weyl spinors from our macroscopic perspective – since this is what we have observed experimentally. In the even-dimensional space, there are two sets of fermions: Corresponding to the usual left-chiral SU(2) doublets is a set of vector-like fermion SU(2) doublets, and corresponding to the usual right-chiral U(1) singlets is a set of vector-like fermion U(1) singlets. These fermions are then expressed as Fourier expansions about the extra-dimensional coordinate(s), which have both even and odd parts (under a parity associated with the higher-dimensional space). The even part contains a zero-mode and a tower of Kaluza-Klein excitations, while the odd part contains just a tower of Kaluza-Klein excitations. This division into even and odd terms is associated with chirality in the effective 3D space: The even part of the SU(2) doublet is interpreted as the left-chiral zero-mode doublet and one tower of Kaluza-Klein excitations, while the odd part of the SU(2) doublet is a second tower of Kaluza-Klein excitations that is right-chiral; the even part of the U(1) singlet is right-chiral, with a zero-mode and a tower of right-chiral Kaluza-Klein excitations, and the odd part of the U(1) singlet is a tower of left-chiral Kaluza-Klein excitations. This way, the zero-modes, which correspond to the usual Standard Model particles, include only the usual SU(2) left-chiral doublets and U(1) right-chiral singlets. There are two towers of Kaluza-Klein excitations associated with the usual Standard Model particles: one corresponding to the usual SU(2) left-chiral doublets and U(1) right-chiral singlets, plus an additional tower with SU(2) right-chiral doublets and U(1) left-chiral singlets.

Suppose that the LHC or another collider produces some heavy particles with a mass of 350 GeV/c^2 or more. This might be ascertained by observing an excess of various final states compared to Standard Model predictions, and analyzing the signal closely to determine the masses of the decaying particles. This would be a signal for some new physics mechanism. So an important question to ask is this: How do we know which form of new physics is being observed? Extra dimensions is not the only anticipated model of physics beyond the Standard Model: There is supersymmetry, Grand Unified Theories, technicolor, and a host of other theories. Thus, we need to not only distinguish signals for new physics from Standard Model backgrounds, but we must also compare signals for different forms of new physics.

Most forms of new physics will result in the production of heavy particles (a couple hundred GeV/c^2 or more) at colliders with high enough beam energies. Either such particles will be observed at the next generation of high-energy colliders, beginning with the LHC, or the bounds on the masses of the heavy particles that they predict will need to be much higher than will be theoretically plausible (there is theoretical motivation that they should not be too much higher than the weak scale of about a TeV, in the case of most mechanisms of new physics). New light particles are a problem – they require explanation for why they have not been observed already. For example, there is no room for a fourth light neutrino in existing collider data. Thus, new particles predicted by theories beyond the Standard Model are generally expected to be heavy, if they exist at all.

The UED model predicts the existence of Kaluza-Klein excitations in the effective 3D theory, with masses of 350 GeV/c^2 or more [T4]. Compared to some signals for new physics, these Kaluza-Klein excitations must be pair-produced in the UED model. So checking the number of particles produced and comparing their masses is one way to see if the new physics may correspond to the UED model. Another feature of the UED model is the two towers of Kaluza-Klein excitations with both left- and right-chiral singlets and doublets if there are an odd number of extra dimensions that are universal (meaning an even number of spatial dimensions all together). While the nature of the spin of the Kaluza-Klein excitations may not be as easy to determine (especially, as they may be easier to produce via the strong interaction, while the nature of their spin is primarily associated with their coupling to the $W^\pm$), the tower will be a smoking gun. For example, if there are Kaluza-Klein excitations that interact like electrons with masses of 500 GeV/c^2, 1000 GeV/c^2, 1500 GeV/c^2, and 2000 GeV/c^2, this regular pattern is much more likely to be explained as a tower of Kaluza-Klein excitations than a group of particles from some other new physics mechanism that happen to have a mass-spacing of 500 GeV/c^2. Seeing multiple forms of the new physics – e.g. through a few different processes, rather than just one form of the signal – will help to determine the source of any new physics observed. Analyzing each signal in detail will also aid in this.

It is also possible to develop non-universal models, although the UED model and graviton-only models may be more aesthetically appealing (and have a simpler structure). In this case, some particles may be confined to the usual 3D space that we observe, while others may propagate into one or more extra dimensions. For example, one such non-universal model that is popular in the literature allows the gauge bosons – the gluons, photon, and $W^\pm$ and Z bosons – to propagate into one extra dimension, whereas the Standard Model fermions – the quarks, charged leptons, and neutrinos – are confined to the usual three macroscopic dimensions. Since this is a sort of hybrid model – a mixture of the graviton-only model and UED model – we might naïvely expect the bound for such extra dimensions to be weaker than it is for gravitons (about a tenth of a millimeter) and stronger than the UED bound (corresponding to a mass greater than GeV/c^2). However, this naïve expectation would be incorrect: This "fermiphobic" model results in the strongest bound of all – a few TeV – from high-energy collider data from Fermilab and LEP.

Here is the reason: It turns out that Kaluza-Klein number is not conserved for interactions involving fermions in the fermiphobic model. Two fermions couple to a single gauge boson, so the Kaluza-Klein excitations of gauge bosons can only couple to fermions by violating Kaluza-Klein number at the tree-level. However, this is okay in the fermiphobic model, unlike the UED model. In the fermiphobic model, the fermions reside on a 3D space, which we call the Standard Model wall. The Standard Model wall absorbs the unbalanced component of momentum along the extra dimension when two fermions interact with a Kaluza-Klein excitation of a gauge boson. This is not possible in the UED model, since in that case the Standard Model wall is 4D (or higher), and so includes the extra dimensions that are universal. Anyway, the violation of Kaluza-Klein number in the fermiphobic model leads to a much larger class of Feynman diagrams, and allows Kaluza-Klein excitations to be produced singly (instead of in pairs) – and even includes processes without Kaluza-Klein excitations in the final state (in this case, they appear in the propagators only). If Kaluza-Klein excitations had a mass of a mere 350 GeV/c^2 in the fermiphobic model, they would have been produced copiously at the Fermilab. Existing collider data shows that Kaluza-Klein excitations must have a mass of several thousand GeV/c^2 for the fermiphobic model.

There may also be theoretical motivations for non-universal models. For example, with six extra dimensions available from superstring theory and six flavors of quarks and leptons – up, down, charm, strange, top, and bottom quarks, plus three types of charged electrons and their associated neutrinos – it is tempting to develop a model where each generation of quarks propagates into a different extra dimension. Here is another example: Dedicate three extra dimensions for the SU(3) strong interaction between quarks, two extra dimensions for the SU(2) interaction for the weak interaction, and one extra dimension for U(1) for the electromagnetic interaction. There is much room for creative model-building, but it is only a philosophic model unless the model makes predictions that can be plausibly tested in a future high-energy collider or other test of particle physics.

In summary, there are three main models in mind for high-energy collider searches for extra dimensions: A model where only gravitons propagate into the bulk, the UED model, and the fermiphobic model. The graviton-only model may feature extra dimensions as large as about a tenth of millimeter, whereas the masses of Kaluza-Klein excitations must be greater than about 350 GeV/c^2 in the UED model or several thousand GeV/c^2 in the fermiphobic model (smaller than the nucleus of an atom).

9.4 Astrophysical and Cosmological Constraints

A collision between particles in a high-energy collider results in new particles in the final state. Similar scattering processes occur in nature, especially in astrophysics or cosmology, where they can occur in very large numbers or very high energies. Kaluza-Klein excitations can affect observations of these interactions just as they can for high-energy collider data.

There are an abundance of interactions occurring in our sun and other stars everyday. For example, the protons of hydrogen atoms bind together to form helium nuclei through a process called nuclear fusion. The sheer number of photons – the electromagnetic energy that we receive from the sun, which provides heat to warm the earth and light for us to see – that the sun supplies everyday illustrates the incredibly large number of processes occurring in the sun and releasing energy; these are processes which produce photons and other particles in the final state. We observe the results of such interactions through telescopes. We can also sort out these signals by intensity and frequency by passing the light through a spectroscope – which consists of a grating or prism that disperses the light (separating it into bands of different wavelength, or color). Every element has a unique signature when viewed through a spectroscope, which allows us to determine the composition of stars. Thus, we are able to determine which processes are occurring in stars, and how often they occur, from astronomical observations. Since Kaluza-Klein excitations can affect these processes, astrophysical observations can provide evidence for or constraints on the size of extra dimensions.

Astrophysical observations turn out to be most relevant for gravitons propagating into the bulk. Processes that occur naturally in stars will produce Kaluza-Klein excitations of gravitons in addition to the ordinary final states. Some energy is radiated away into the extra dimensions through the production of Kaluza-Klein excitations of gravitons. This also affects the thermodynamics – i.e. the temperature the star and the rate at which it changes over time.

The chief astrophysical constraint comes from observations of supernovae, owing to the extreme number of processes involved that release enough energy to increase the brightness of the star more than a billion-fold. Red giants also produce a great deal of energy. The sun also merits close examination since we receive the bulk of our energy from it owing to its close proximity. In particular, Supernova SN 1987 A provides the strongest astrophysical constraints. Data from SN 1987 A require that the electroweak scale exceed about 10 TeV unless there are three or more extra dimensions, in which case the constraints are much more relaxed [T2].

It is also important to consider very high-energy cosmic rays, which carry up energies upwards of 10^{20} eV $= 10^8$ TeV on their journey through space and collide with particles in earth's upper atmosphere. The collision energy can correspond to a proton-proton collider with a beam energy of 100-1000 TeV – compare with the LHC, which will run at 14 TeV. Although these are higher energies than we can produce in the laboratory, they do not lead to constraints as stringent as colliders provide. This is in part due to the fact that high-energy cosmic ray experiments are much less controllable: In a collider, we control when and where we will have a collision, and can produce a very large number of similar collisions; for high-energy cosmic rays, we have to set up a detector and wait for whatever events may come. Cosmic rays do not offer as good a means of probing for extra dimensions as high-energy colliders, yet they do provide a means of observing events that occur at energies above the effective fundamental scale – the single unifying scale of the electroweak and gravitational interactions on the order of a TeV. However, they do not probe short distances as well as high-energy colliders, even though the collisions can occur at much higher energies in cosmic rays, due to the nature of the collisions (they are diffractive and lose energies through cascades of electromagnetic showers).

In addition to astrophysical constraints on the size of extra dimensions, there are bounds arising from cosmological observations. One especially important cosmological issue is Big Bang nucleosynthesis, since a small change (as little as 10%) in the rate of expansion of the universe during this very early stage could have a very significant impact on modern-day observations. As the rate of expansion is related to temperature, thermodynamic considerations must be taken into account in models of the Big Bang. Kaluza-Klein excitations of gravitons radiating energy into the bulk affects the expansion rate. Constraints arising from Big Bang nucleosynthesis can push the fundamental scale up to 10 TeV or higher for two large extra dimensions, or up to a few hundred MeV for up to six large extra dimensions [T2].

The UED model features an additional cosmological issue: There will be one or more stable Kaluza-Klein excitations of Standard Model particles due to conservation of Kaluza-Klein number. This means that there must be an abundance of lightest Kaluza-Klein excitations in our universe – since whenever the lightest Kaluza-Klein excitations are produced, they do not decay to other particles. This could pose a cosmological problem if there are stable Kaluza-Klein excitations of leptons or quarks during Big Bang nucleosynthesis. However, the lightest Kaluza-Klein excitation could very well turn out to be the photon, which is electrically neutral and has zero rest-mass.

Further Reading:

There is an abundance of published research on the phenomenological searches for large extra dimensions, including tests of gravity, collider searches, astrophysical searches, cosmological searches, and other searches. The vast majority of these papers are highly technical, designed for specialists in particle physics, field theory, string theory, and related branches of mathematics and physics. As this book is intended as an enjoyable for a more general mathematically-minded audience, the technical papers have been kept to a minimum. If you are interested in these technical works, the references suggested in Chapter 8 [T2-T4] are a good place to start. If you are curious about tests of Newton's law of gravitation, see [T5]. The current status of the Standard Model of elementary particles and their interactions is professionally and exhaustively tabulated at [T6]. The technical papers on the subject of large extra dimensions are available at online at www.slac.stanford.edu/spires/find/hep. Readers looking for more accessible articles in this area should try [A1]. For an accessible introduction to elementary particles and collider physics, see [A2].

10 Spacetime

In this last chapter, we consider the role of time as an integral component of the spacetime continuum in Einstein's special and general theories of relativity. We also probe the main concepts underlying quantum mechanics. The theories of relativity and quantum mechanics are integral components of quantum field theories, which form a mathematical description of the elementary particles of our universe and their interactions. However, we will limit ourselves to a conceptual understanding of a few ingredients of quantum field theory. We will discuss the ongoing quest for a theory of everything, and discuss the basic elements of string theory, superstring theory, and M theory.

10.0 Special Relativity

You have probably heard the phrase, "It's all relative," – and it is, but there is much more to Einstein's theory of relativity than this simple notion. There's a similar principle of things being relative in Galilean relativity. However, observations of high-speed objects agree with Einstein's special relativity, and not Galilean relativity. Let us first explain what *it* is and how *it's relative* in the intuitive theory of Galilean relativity, and then we will tackle the basics of special relativity.

Suppose that you are standing at a train station, close to the tracks, and a train passes by with a constant velocity of 5 m/s to the north. You see a man on the train walking north with constant velocity. A passenger on the train also observes the man walking. The man gets 3 m further from the passenger every second, so the passenger measures the man's speed to be 3 m/s. This is the man's speed relative to the passenger. Relative to you, however, the man is moving 8 m/s because each second the train gets 5 m further from you and the man gets another 3 m further from you on top of this. So, *velocity* is *relative*, and common human experience suggests that relative velocity is simply found by adding the velocity vectors together. This intuitive notion is called Galilean relativity.

Puzzle 10.1: A boy rides in a boat with a velocity of 40 mph to the north, while a girl rides in a boat with a velocity of 20 mph to the south, going away from the boy, where these velocities are measured relative to the lake. If the boy throws an orange 10 mph to the south, relative to his boat, what is the orange's velocity relative to the girl?

However, while Galilean relativity agrees with our experience with low-speed observations, it does not agree with measurements of objects with speeds that are a significant fraction of the speed of light – or the relativity of light itself. According to Galilean relativity, and thus our intuition from everyday experience in a world of low speeds, if a train is moving half the speed of light, or $0.5c$, and a man on the train turns on a flashlight, the light emanating from the flashlight should travel the speed of light, c, relative to the train, and a speed of $1.5c$ relative to the train station. This expectation disagrees with experiment, which shows that the speed of light is independent of the speed of the source or observer.

Counterintuitively, experiment shows that light travels approximately 300,000,000 m/s relative to the train, as expected, but also 300,000,000 m/s relative to the station. Think about this for a while. The train is moving half the speed of light, so this unrealistic high-speed train (conceivably, more realistic for a futuristic spaceship) travels 150,000,000 m each second relative to the train station. According to a passenger on the train measuring the speed of light emitted from the flashlight, the photons travel 300,000,000 m further than the train each second. An observer at the train station measures the train to get 150,000,000 m further from the station each second, yet still observes the photons to travel a mere 300,000,000 m further from the station each second – not the expected 450,000,000 m.

The same photons emitted by the flashlight that get 300,000,000 m further from the train station each second also get 300,000,000 m further from the train each second – well, that's the way it seems at first, which seems paradoxical. We're comparing observations made by different observers, not the same observer. The passenger in the train observes, relative to him, that the photons travel 300,000,000 m/s, while the observer at the train station finds that the photons travel 300,000,000 m/s relative to the station. These measurements are made by different observers moving at different speeds (or not moving at all). Relative to any observer, light will travel 300,000,000 m/s, regardless of the motion of the source or observer. The speed of light is a universal constant.[10]

---

[10] There are cosmological observations that may be explained if the speed of light is changing slightly over very long periods of time. However, this does not impact our current discussion unless we

The constancy of the speed of light and, hence, evidence that Galilean relativity does not apply to very high-speed objects was first shown by Michelson and Morley in 1887. Ironically, much the opposite was anticipated, but they published their results even though their experiment may have seemed a failure in light of what were common expectations at that time in the history of physics. Now their experiment is famous, and we know that it was not a failure, but a triumphant discovery in the development of physics. They used an interferometer, as illustrated below. A partially reflecting mirror splits an incident beam of light into two paths that travel at right angles. Each of the split beams reflects off a mirror. Light from each arm of the interferometer joins up once again at the partially reflecting mirror, which heads to a telescope. An interference pattern results on the screen, depending on the phase difference between the two beams.

Light is an electromagnetic wave – electric and magnetic fields that oscillate as the beam of light propagates. Light from the two beams is in phase when the oscillation of their electric and magnetic fields is synchronized. If, for example, one beam spends more time traveling along its arm of the interferometer, the two beams may not be in phase when they reach the telescope.

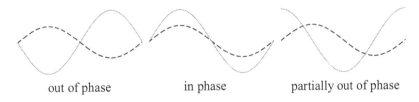

out of phase   in phase   partially out of phase

As the earth travels through space, light traveling along one arm of the Michelson-Morley interferometer may be traveling in the same direction as the earth, for example, while light in the other arm would be traveling perpendicular to the earth's velocity.[11] According

---

consider measurements of events that may last for a very large number of years.

[11] You might think to measure the earth's velocity relative to the sun, but the sun itself is traveling through our galaxy. You could measure the earth's velocity relative to the center of the galaxy, but then the galaxy is moving relative to other objects in the universe. In this Galilean model, there was believed to be some preferred reference frame, called the luminiferous ether. We'll just call it earth's velocity, but in this picture it is earth's velocity relative to the ether.

to Galilean relativity, the earth's velocity should affect the velocity of light in the two beams. When one beam is aligned with the earth's velocity, one of the two beams will take less time to return to the mirror and the two beams will be significantly out of phase. Michelson and Morley rotated their apparatus through 360° to see how this affected the interference pattern observed through the telescope. For some angles, they expected the two beams to be in phase (when each beam makes a 45° angle with earth's velocity), while for other angles, to be significantly out of phase (namely, when one beam is parallel to earth's velocity). However, they observed no significant difference in interference patterns as they rotated the beam, which demonstrated that light travels the same speed in all directions relative to earth's velocity.

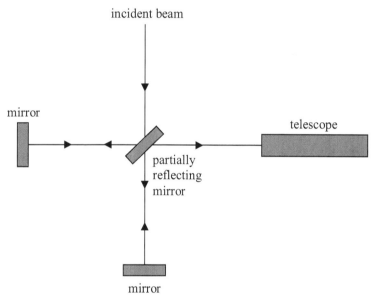

the essence of the Michelson-Morley interferometer: a partially reflecting mirror splits an incident beam in two; each beam reflects from a mirror and returns to the partially reflecting mirror; light entering the telescope is the superposition of light from each of the two mirrors; as the entire apparatus is rotated, the interference pattern seen through the telescope would change if measurement of the velocity of light is governed by Galilean relativity

Puzzle 10.2: Consider the illustration of the Michelson-Morley interferometer. Determine which beam of light returns to the partially reflecting mirror in the least time if the earth's velocity is directed (A) upward, (B) to the left, and (C) diagonally upward and left in the diagram.

Another way of interpreting the results of the Michelson-Morley experiment is that there is no preferred, or absolute, reference frame. The speed of light is measured to be the same value regardless of the motion of the source or the observer. In 1887, light was thought to travel in a medium, called the ether. Since light is a wave, and all other waves known at the time required a medium in which to propagate – e.g. sound, consisting of variations in pressure, requires a medium like air or wood, and so cannot travel through a vacuum – the ether was the hypothesized medium of light. The ether also constituted an absolute frame of reference in the ether theory. However, we now know that the ether does not exist – light can travel through a perfect vacuum – and there is no such thing as an absolute reference frame. As we shall see, no inertial observer has preference over any other.

Historically, it took time to check, accept, sort out, and understand the results of the Michelson-Morley experiment. Albert Einstein's special theory of relativity provides a simple explanation – well, simple in that the theory derives from two fundamental postulates, but not so simple in that it is conceptually challenging, involving abstract notions of spacetime and high-speed travel with predictions that defy our experience with low-speed observations that appear to agree with Galilean relativity. The two postulates of Einstein's special theory of relativity are:

> 1. The laws of physics apply equally to all inertial reference frames.
> 2. All inertial observers measure the same value for the speed of light in vacuum, 300,000,000 m/s, regardless of the motion of the source or the observer.

The invariance of the speed of light observed by Michelson and Morley is taken as the second fundamental postulate. The first postulate states that it is not necessary to sacrifice or modify the laws of physics in order to accommodate the observed constancy of the speed of light. (There were other proposals that favored revising the laws of physics; however, experiment confirms Einstein's theory, and not the other proposals.) A corollary of the first postulate is that there is no absolute or preferred reference frame – any inertial observer is entitled to say, "I'm at rest," and be correct. Since the laws of physics apply to all inertial reference frames equally, there is no experiment that can be done to distinguish between being at rest and moving with constant velocity. In this sense, *it's all relative*.

Both postulates involve inertial observers. Einstein's special theory of relativity applies to inertial observers; his general theory of relativity is needed for non-inertial observers. Recall, from Chapter 7, that inertia is a natural tendency of an object to travel with constant momentum, and that momentum equals mass times velocity. As such, an inertial observer is one who travels with constant velocity, assuming the observer to have constant mass (not to be confused with the relativity of mass, to be addressed toward the end of this section).

An inertial observer travels with constant velocity.

Interestingly, the two simple postulates of special relativity lead to a spacetime continuum with some very counterintuitive features. In order for the speed of light to be the same for all inertial observers and for the laws of physics to hold in all inertial reference frames, measurements of time and length cannot be the same for all observers. To see this, return to our earlier consideration of a train (or rocket) traveling half the speed of light, where we compared measurements of the speed of light emitted by a flashlight onboard the train relative to observers on the train and at the station. If both are to agree that the light travels 300,000,000 m ahead of themselves each second, without modifying the laws of physics for inertial observers, measurements of length and time must be different for different observers in order for the speed of light to be invariant.

Since our ordinary notion of spacetime is evidently different in special relativity, Einstein developed a simple conceptual clock to allow for time comparisons, called a light clock. A light clock consists of two parallel mirrors with a ray of light bouncing back and forth between the two mirrors (perpendicular to each mirror). The time it takes for light to travel from one mirror to the other defines a basic unit of time. Since all inertial observers measure the speed of light to be the same value, this light clock is a very useful tool in special relativity.

Consider an astronaut traveling in a rocket with a constant speed of v, which could be $c/2$, for example, which is a shorthand way of writing 150,000,000 m/s since c is the speed of light (300,000,000 m/s), relative to the earth (or any inertial observer). The astronaut has a light clock on board the ship, and according to this light clock time is normal on the ship. An observer on the earth also carries a light clock, and time on earth is normal according to the earth's light clock. However, time appears to run differently when one observer measures time with a light clock that is in relative motion with the observer – e.g. if the observer on earth compares the light clock on the rocket to the light clock on earth, the two light clocks will appear to not be synchronized, even if (as we shall assume) they were synchronized before the rocket took off.

Let us consider the light clock on the rocket from both points of view. The astronaut is entitled to think she is at rest. From her point of view, the light on her light clock travels perpendicular to its mirrors. She measures the time t' it takes for light to travel the distance d' between the mirrors to compute the speed of light: $c = d'/t'$. Since the rocket is moving relative to the earth, the light on the rocket's light clock travels a distance greater than d' from the earth's vantage point. This is because the mirrors, which are on the fast-moving rocket, are moving relative to the earth. Thus, for the observer on earth to measure the same speed of light for the rocket's light clock, in accordance with the second postulate, the time it takes for light to travel from one mirror to another must be greater than t' from the earth's point of view: $c = d/t$, where d is the distance light travels in going from one mirror to another relative to the earth and t is the time it takes for light to make this trip relative to the earth. A larger numerator requires an proportionally larger denominator in order for the two fractions, d/t and d'/t', to both equal the same value c. Mathematically, as worked out below, $t = \gamma t'$, where γ is the time dilation factor.

Conceptually, we interpret this result as follows: Moving clocks run slower. All inertial reference frames are created equal – i.e. if you are moving with constant velocity, you have equal right to believe that you are at rest (as equal as any other object moving with constant velocity). For any object moving relative to you, time runs slowly relative to you. Suppose you watch a movie on a television that is at rest relative to you and the movie ends in two hours. If you were to watch the same movie on a television that was moving very fast relative to you, it would take you more than two hours to watch the same movie.[12] However, if the television were on a fast-moving rocket, an astronaut on the rocket would watch the movie in two hours, since the television would be at rest relative to the astronaut.

---

[12] In case you are wondering about the time it takes for light to reach you from the moving television, even if you account for this relay time, you will still find that time is dilated.

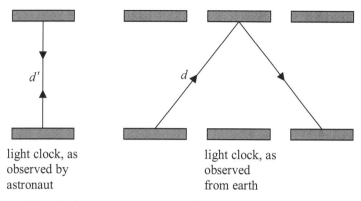

light clock, as observed by astronaut

light clock, as observed from earth

a light clock on a rocket as viewed from the perspective of the astronaut and an observer on earth; since the speed of light must be the same for both observers and since light travels a longer distance from the earth observer's perspective, time must be dilated for the astronaut relative to earth

$$c = \frac{d'}{t'} = \frac{d}{t} = \frac{\sqrt{d'^2 + v^2 t^2}}{t}$$
$$c^2 t^2 = d'^2 + v^2 t^2$$
$$c^2 t^2 = c^2 t'^2 + v^2 t^2$$
$$t = \frac{c}{\sqrt{c^2 - v^2}} t' = \gamma t'$$

Puzzle 10.3: Provided that she has a fast enough spaceship, could a mother go on a journey through space and be younger than her children when she returns? If so, could she possibly return before her children were born?

Puzzle 10.4: Without living more than 100 years, how could a human go an a journey through space and return in the year 10,000 A.D.? Could a human return in the year 10,000 B.C.?

One consequence of the two postulates of special relativity is that time is dilated for objects in relative motion. On a similar note, observers in relative motion will not agree on simultaneity. That is, if two events are simultaneous in one inertial reference frame, they may not be simultaneous in other inertial reference frames. Since no reference frame has a preference over any other, the notion of two events occurring simultaneously is purely relative.

Puzzle 10.5: Imagine that a cloning process is developed to clone a chimpanzee that successfully always produces identical clones, which all live for exactly 50 years relative to their own light clocks. Further, two such clones are produced in spaceships as the spaceships pass one another in opposite directions. This way, the cloning scientists are sure that the clones are born *at the same time*, since they are right next to one another (momentarily) at birth. Relative to each chimpanzee, each will find that the other chimpanzee ages more slowly if they make precision relativistic measurements (even if they subtract the time it takes for light to reach their spaceship). According to Einstein, both are correct! Surely, if they meet face to face at some later time, then they still can't both be correct in thinking that the other is younger. Resolve this paradox.

In addition to time dilation, there is length contraction: Moving objects appear shorter along the direction of relative motion. For example, a rocket moving at a high speed away from the earth appears shorter, but not taller or wider. To see how length contraction can come about, consider making a journey to Alpha Centauri, which is about 4 light-years away (so it takes light 4 years to get there). If you could travel $c/2$ relative to our sun, it would take 8 years to get there – relative to an observer on earth. However, traveling on the spaceship, less than 8 years would pass on your spaceship's light clock as a result of time dilation. You have to be careful here: Since time passes more slowly on the spaceship that is in motion relative to the earth, less time passes on the spaceship's light clock than on the earth's light clock (where all measurements are made from the earth). So how can the spaceship, with a speed of $c/2$, reach a star that is 4 light-years away in less than 8 years – i.e. according to the earth-based observers, the pilot and passengers on the spaceship age less than 8 years, so they will see fewer than 8 years pass on their calendar when they arrive at Alpha Centauri.

The answer is length contraction. Let's look at this from the spaceship's point of view now. Passengers on the spaceship are free to say that they are at rest, while Alpha Centauri is approaching their spaceship with a speed of $c/2$; after all, it's all relative. The space travelers reach Alpha Centauri in less than 8 years, as we have already seen from earth's vantage point. Since Alpha Centauri is traveling $c/2$, for Alpha Centauri to reach them in less than 8 years, the space travelers naturally conclude that length has contracted – that is, the sun and star that are in motion relative to the spaceship are closer than 4 light-years apart due to length contraction. The formula for length contraction is $d = d'/\gamma$. Ultimately, the relative differences in time and length come about in order to preserve the same speed of light for all observers without sacrificing the laws of physics – exactly, the two postulates of special relativity.

Puzzle 10.6: An airport hanger has two doors – one on the north side and one on the south side. The doors can open or close (virtually) instantaneously. The distance between the two doors is 30 m. When a 40-m long spaceship is parked inside, at least one door has to be open because 10 m of the spaceship has to stick out. Now the spaceship takes off and returns, approaching the hanger at nearly light-speed. An automated controller (with virtually zero reaction time) opens the south door so that the spaceship can enter and closes the south door immediately behind the spaceship, and also opens the north door just in time. A repairman inside the hanger notices that for a moment, the spaceship is inside the hanger while both doors are simultaneously closed, yet neither door has been damaged. How is this possible?

The predictions of special relativity, including time dilation and length contraction, have been successfully tested in various ways. For example, muons are particles very similar to electrons except for having more mass. Muons are constantly produced when high-energy cosmic rays from space collide with atoms in earth's atmosphere. They are produced at high altitudes and decay very quickly to electrons and other particles. We can also produce muons in the laboratory, so they are well understood. Based on their average lifetime (i.e. how long they exist before decaying) and their altitude when they are produced by high-energy cosmic rays, almost none of the muons should reach earth's surface – they have too far to travel and don't live that long. Nonetheless, detectors measure that a significant number of muons do, in fact, reach the ground – far too many to be attributed to an experimental source of error. However, when we realize that these muons travel at very high speeds – close to the speed of light – and therefore experience time dilation, we can explain the number of observed muons with special relativity: The muons live longer than expected – long enough for many of them to reach the ground. In fact, the theoretical prediction, after applying special relativity, agrees very well with the experimental observation.

Puzzle 10.7: From the perspective of the muons, they live normally – i.e. they do not experience time dilation relative to themselves. From their perspective, how do you account for the fact that many muons are able to reach earth's surface?

In addition to time dilation and length contraction, mass is also relative. Consider the following simple argument for why the mass of a moving object should increase as it approaches the speed of light. According to Newton's second law, the net force acting on an object that has constant mass equals the object's mass times its acceleration: $\vec{F}_{net} = m\vec{a}$. Suppose that you apply a meager 1-N force to a 1-kg mass: In this case, you get an acceleration of 1 m/s^2. This means that, for as long as the 1-N force is applied, the object will gain 1 m/s of speed each second. After 10 s, its speed will be 10 m/s; after 1 hr, its speed will be 3600 m/s; and so on. If you continue to apply this 1-N force to the same object for more than 300,000,000 s, the object will be traveling faster than light!

This is a long time to wait, of course. Plus, it's not as easy to apply a 1-N force as it might first seem. You can't just push a box, for example, because then you'll soon have to move very rapidly yourself just to keep up with the box. However, we can accelerate very light particles, like the electron, to very nearly lightspeed very easily in the laboratory through a similar method. Since the electron has such a tiny mass, it is very easy to achieve very large accelerations – so we can accelerate it to nearly light speed very rapidly (unlike the 300,000,000 s of our previous example). Particle accelerators use an electric field to do this. However, unlike our previous example, although we can accelerate electrons to very high speeds very easily, it becomes increasingly difficult to accelerate the electron as its speed gets closer and closer to the speed of light. The reason is that the electron's mass increases as it approaches the speed of light.

So, returning to our 1-kg mass, if we apply a 1-N force to it, it will gain 1 m/s of speed each second until its speed is close to the speed of light. Once it has a very high speed, the mass will no longer be 1 kg: The mass of the object will increase according to $m' = \gamma m$. The closer its speed gets to the speed of light, the more mass the object has.

As the speed of an object approaches the speed of light, its mass becomes infinite, time slows to a standstill, and its length shrinks to zero. But since the mass of the object grows very rapidly as its speed gets very close to the speed of light, it would take an infinite force to actually reach the speed of light. So any object originally traveling under the speed of light is apparently condemned to always travel slower than light. If you're wondering what happens if you travel faster than light – i.e. if time reverses, it would be the basis for time travel – you first need to figure out how you're going to get an object to travel faster than light. There are hypothetical particles, called tachyons, which have been proposed to travel faster than light – but if they travel faster than light when they are produced (in a collision of some kind, for example), then we don't have the issue of how to jump abruptly from slower than light to faster than light. If you're a science fiction writer, perhaps you can exchange all of the particles making up your protagonist for tachyons, temporarily, and then after going back in time re-exchange them – at a possible philosophical cost, since you have to wonder if this is the "same" person after the exchange. I suppose if you're reading about extra dimensions, time travel shouldn't seem too far-fetched…

In the other extreme, at low speeds, γ is approximately equal to one. Our everyday experience involves low speeds, since we never travel anywhere remotely close to light-speed. At low speeds, we don't observe the strange effects of special relativity. Special relativity becomes Galilean relativity at low speeds.[13] There is no time dilation, length contraction, or mass increase – well, there is, but it's extremely small, so much so that we wouldn't observe it without very high precision equipment – at low speed travel. Thus, our intuition is very Galilean, and we tend to be reluctant to accept the consequences of special relativity, which seem contrary to our low-speed experiences. If you realize that we have no experience at high speeds, and think scientifically to base judgment on experiment, then we may accept the scientific results that confirm the effects of time dilation, length contraction, and mass increase for speeds close to the speed of light (as measured by an observer who deems himself/herself to be at *rest* – as all inertial observers are free to do, without any contradiction).

We conclude this section with a note for the mathematically-minded reader. The equations for time dilation, $t = \gamma t'$, length contraction, $d = d'/\gamma$, and mass increase, $m' = \gamma m$, are very easy

---

[13] Obviously, measurement of the speed of light is inherently a special relativity problem, since light can't travel at slow speeds; Galilean relativity simply cannot reconcile the observed invariance of the speed of light.

misuse if not properly understood. For example, if you confuse which observer measures t with which observer measures t', the math will tell you that time is contracted instead of dilated. This is no fault of the math, of course – just like a computer, if you feed garbage into an equation, you get garbage out of it. How to properly use the equations comes from a good conceptual understanding, beginning with precise definitions of the fundamental concepts. The concepts of *proper time* and *proper length* are useful for determining which variable should be primed or unprimed in the formulas. If an observer sees two events occur at the same place, this observer is said to measure proper time, which is denoted as t'. Compare with our time dilation derivation, where the rocket observer always sees his own light lock in the same place, whereas the rocket's light clock was moving relative to earth: Hence, we used t' for time measured on the rocket and t for time measured by earth. Similarly, proper length is defined for an observer for which the object appears at rest, which is denoted as d'. For example, for muons heading toward earth's surface, from the perspective of the muons time is normal so they experience an ordinary lifetime, but reach the surface because length is contracted relative to them (in their frame, they say that they are at rest, while the earth moves rapidly toward them). Here, d' is the distance to earth as measured on earth, while the muons measure a shorter distance d. Compare to the earth-based perspective, where the distance measured by earth is correct, but the muons reach the earth because their lifetime is dilated. The two perspectives are each correct in their respective reference frames, and both agree on the number of muons that reach earth's surface. That's the principle of relativity – every inertial reference frame is as good as any other. The adjective *proper* is not used to give preference to any reference frame – rather, its purpose lies in its convenience to help determine which set of variables to use for which reference frame in the equations as they have been provided.

10.1 General Relativity

Special relativity shows how our usual perception of spacetime is altered regarding measurements of objects moving nearly light-speed – namely, moving objects appear shorter than they do when they are at rest, time runs slower for moving objects, and moving objects appear to have more mass. This unusual – from our daily experience with slow-moving objects – spacetime continuum follows directly from the two postulates of special relativity: In order to explain the Michelson-

Morley result that all observers measure the same speed of light regardless of the motion of the observer or the source while also insisting that the laws of physics are equally valid in any inertial reference frame, length must contract and time must dilate for objects moving relative to inertial observers. In this way, space and time are integral components of a spacetime continuum – they balance out in order to preserve the invariance of the speed of light.

Special relativity has very limited application, though, since it deals only with inertial observers. Recall the resolution of the twin paradox, in which one twin must accelerate in order for the two twins to make face-to-face comparisons. Special relativity deals with objects moving with constant velocity, and does not account for the spacetime effects of acceleration. However, many objects accelerate. For example, you accelerate when you get in a parked car and go from 0 to 60 mph in 10 seconds, and a rock accelerates when you throw it through the air. Also, a simple change in direction, even with constant speed, results in acceleration – so special relativity deals exclusively with objects traveling with constant speed in straight lines. Einstein's general theory of relativity extends his theory of special relativity to account for acceleration. The theory of general relativity is inherently much more mathematically complex – but, of course, we will focus on some of the main conceptual features of the theory.

General relativity, like special relativity, features a principle of equivalence. In special relativity, the main underlying equivalence principle is that any inertial reference frame is as good as any other – i.e. there is no preferred or absolute reference frame, or the laws of physics apply equally to all inertial reference frames. The main underlying equivalence principle for general relativity is this: Uniformly accelerated motion is equivalent to being at rest in a uniform gravitational field.

Each equivalence principle can be illustrated by imagining a scientist in a laboratory that is completely sealed off from the outside environment. This means that there are no windows to see outside, no sounds from outside can penetrate the walls, etc. The only way the scientist can conceivably learn anything about what is outside the box (the laboratory) is by performing experiments inside the laboratory. According to special relativity, if the laboratory is moving with constant velocity, there is no experiment that the scientist can perform to determine whether or not the laboratory is at rest or moving. This follows since the laws of physics are the same, from the first postulate, in any inertial reference frame – moving or not. According to general relativity, the laboratory could either be (1) at rest on the surface of a planet or (2) moving with uniform acceleration in a region of space that is free of gravity, and there is no experiment that the scientist can do to determine whether the laboratory is in state (1) or state (2).

Let us illustrate these equivalence principles by considering a couple of examples. Imagine such a laboratory that is inside a spaceship in deep space, far from any measurable gravitational fields (i.e. gravity is too weak to be noticed). First, assume that the spaceship is not accelerating, which means that it could either be at rest or moving with constant speed in a straight line. This corresponds to special relativity. The scientist can't look outside, hear vibrations, and so on, to see if the laboratory is moving, since this hypothetical laboratory is completely isolated from its environment. There is no gravity, so everything floats weightlessly.[14]

The reason that the scientist can't tell that the spaceship is moving has to do with inertia – objects have a natural tendency to maintain constant velocity. The objects do not fall to the back of the spaceship – already in motion, they continue to remain in motion. From the postulates of special relativity, no experiment will reveal whether or not the spaceship is at rest or moving with constant velocity, since the laws of physics and the measured speed of light would be the same in either case. But if the spaceship changes velocity, that's a different story.

So now let us imagine that the spaceship accelerates – in particular, by gaining speed, while still continuing along a straight line. This corresponds to general relativity. It may eject some steam, for example, in order to achieve this, but the scientist in the laboratory will not be aware of this. What the scientist will observe is that when the spaceship gains speed, everything does fall to the back of the ship. The scientist will be on the back wall of the ship, and it will feel like this wall is the ground. The scientist will experience gravity equal to the acceleration of the spaceship, even though the spaceship is in a gravity-free region of space. Sitting, standing, walking, etc. at the back of this spaceship – which will seem more like a bottom than a back from a passenger's point of view – will be no different than performing these same activities near the surface of a planet. If the scientist picks an object up and releases it from rest, it will fall downward; if he measures

---

[14] You can also experience weightlessness in a region of space where gravity is quite strong, such as is the case for a free-fall orbit – like a satellite circling the earth or an elevator with a clipped cable. However, let's keep our present consideration simple: For our equivalence principle in general relativity, we are presently concerned either with being at rest in a gravitational field or moving with constant acceleration in a gravity-free region.

its acceleration, he will find that it uniformly accelerates. If the acceleration of the spaceship is 9.81 m/s^2, he may (incorrectly!) expect that the spaceship is parked on the surface of the earth. If he throws an object, it will curve downward in a parabola just as it would on earth. There is no way to determine whether or not the spaceship is traveling with constant acceleration is a gravity-free region of space or if it is parked on the surface of a planet – either way, the laws of physics are equivalent.

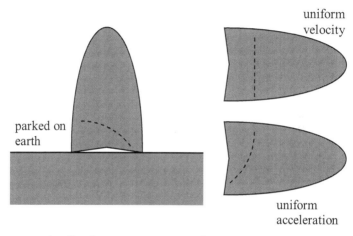

a projectile thrown across a rocket appears to follow a parabolic arc relative to an observer on the rocket if it is at rest in a gravitational field or uniformly accelerating in a force-free region of space; however, if it travels with uniform velocity through a force-free region of space, the observed trajectory is then linear

Now let us imagine that the scientist turns on a laser beam in his laboratory onboard this spaceship. While the speed of light is observed to be a universal constant, which does not depend upon either the motion of the source or the observer, the velocity of light entails some fascinating subtleties. The distinction is that velocity includes direction, whereas speed does not. It turns out that the path of the laser beam does depend upon whether or not the spaceship is accelerating, which the scientist could measure, in principle, though he would need extremely high-precision equipment as the effect is very, very slight except for very high accelerations (for which the scientist will need to be a robot in order to survive).

If the spaceship travels with constant velocity (still in an effectively gravity-free zone), the contents of the spaceship float weightlessly, and if the scientist (also floating weightlessly, or perhaps holding a handle on a wall or strapped in a seat) throws an object, it will travel in a straight line with constant velocity – regardless of the direction in which he throws it. All objects in the spaceship want to continue moving with constant velocity – the same as the spaceship – owing to their inertia. If an object is thrown crosswise (perpendicular to the spaceship's velocity), it does not fall backward, but travels in a straight line across the spaceship (for the same reason that if you throw a pencil straight upward in an airplane, it does not fall back and hit the person behind you, but travels straight up and down back to your hand – the difference being that it falls back down on the airplane due to gravity; on the spaceship, in the present example it would just keep going). If this were not so, you would be able to distinguish between being at rest and moving with constant velocity. Whatever laws of physics you observe in a spaceship that is at rest, you must observe exactly the same laws when the spaceship travels with constant velocity – in accordance with the equivalence principle in special relativity.

Similarly, light has inertia. If you turn on a laser that is at rest, light travels in a straight line along the axis of the laser. The same must be observed if a laser is turned on in a spaceship that is moving with constant velocity, in order to satisfy the equivalence principle. So if the spaceship is traveling $0.5c$ and a laser is turned on perpendicular to the spaceship's velocity (notice that this is different to the flashlight of Section 10.1, where the flashlight was turned on in the direction of travel), the laser beam will not trail backward, but will cut straight across the spaceship, perpendicular to the direction of travel.

If the spaceship accelerates, the laser beam curves. This is once again due to inertia, and relates to the equivalence principle in general relativity. An object thrown in the spaceship curves if the spaceship accelerates in zero-gravity or if the spaceship is parked on the surface of a planet. Light curves in either case for the same reason, except that the curvature of light is extremely slight in comparison since it has such an incredibly large velocity compared to an object that might be thrown. Near the earth's surface, an object thrown horizontally falls downward 4.9 m in one second. Light would also fall 4.9 m downward, if not for the fact that light would be 300,000,000 m from its starting point in one second (light has escape velocity – if you "threw" a rock this fast, it would never hit the ground!); light still curves, but not enough to notice a laser beam bending near earth's surface. Light curves on an accelerating spaceship, which we can understand due to its inertia, so it must also curve due to the presence of a gravitational field, in order to satisfy the equivalence principle of general relativity.

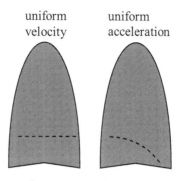

a laser beam travels across a rocket in a gravity-free region of space: if the rocket has constant velocity, the laser travels straight across, but if the rocket accelerates, the laser beam appears to curve (the path shown above is quite exaggerated)

Thus, a fascinating feature of general relativity is that gravity bends light. However, it takes an extreme gravitational field and precision measurements to observe this – like light passing very close to a star. Since light defines our notion of straightness, we interpret this effect as spacetime curvature – i.e. we say that the spacetime is warped, not that light curves; the two perspectives are equivalent. General relativity is mathematically expressed as a four-dimensional spacetime – three coordinates of space and one of time (not counting any extra dimensions) – in which the equivalence principle is expressed geometrically as a spacetime curvature. Conceptually, from special relativity we can see that space and time are intrinsically related: In relativistic measurements, space and time affect one another as they preserve the invariance of the speed of light and provide the same laws of physics to all inertial observers, resulting in the effects of length contraction and time dilation. The sense of spacetime is even more profound in general relativity in its geometric curvature.

The path a point-like object makes as it travels through spacetime is called its world line, tracing out its journey from its past to the present. The countless point-particles (or strings, perhaps) in our universe weave a web of world lines through the spacetime of the cosmos. The future of an object is not "wide-open," but is constrained by a light hypercone (often called a light cone, but with three

179

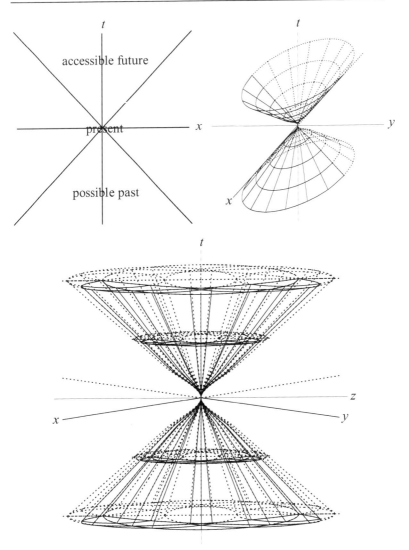

light cones in (1+1)-D, (2+1)-D, and (3+1)-D spacetime: the present lies at the origin, the top of the cone represents points in the future that are accessible without exceeding the speed of light, and the bottom shows possible locations in the past that could have led to the present without exceeding the speed of light; the surface of the cone itself corresponds to traveling the speed of light, the interior represents subluminal velocities, and the exterior is forbidden dimensions of space and one dimension of time, it is a hypersurface bounding a 4D hypervolume, and so strictly is a hypercone). The hypersurface of the light hypercone represents regions of spacetime accessible to a particle that travels the speed of light. Particles with non-zero rest-mass, which necessarily travel less than the speed of light (excepting hypothetical particles called tachyons), can only reach destinations in spacetime inside the hypercone. The region outside the hypercone is strictly forbidden to any particle incapable of achieving superluminal velocities.

The light hypercone arises under the assumption that light travels 300,000,000 m/s in a straight line, but as we have seen, light bends when it passes near a strong gravitational field – i.e. when it passes nearby an object with an extreme amount of mass, such as a star. So, strictly speaking, this "hypercone," which represents the possible future continuations of the world line of a particle, may not be a perfect hypercone, but may be warped: Where a possible future path of the particle would entail passing near a star or black hole, light would not travel in a straight line, but would curve, and this curvature must be reflected in the light "hypercone" if it is to correctly depict the boundaries of the possible continuations of the world line of a particle. Let us look at the gravitational deflection of light for an actual path, instead of looking at the infinite number of possible paths depicted by the interior of the light hypercone.

A star has enough mass to noticeably warp spacetime – well, with precise instrumentation, anyway; not visibly to the naked eye. We normally think of the space of the universe as a 3D Euclidean space. One feature of a Euclidean space is that the shortest distance between two points is a straight line. However, spacetime curvature is a non-Euclidean space, due to the warping of spacetime caused by very massive objects. Imagine that the space of the universe is 2D, and for the most part appears like a plane. You can model such a universe with a pillow to see the simplest effect of spacetime curvature: Place a heavy ball on the pillow. You will see that the weight of the ball bends the region of space (the surface of the pillow) in the vicinity of the ball, while the remainder of the pillow appears relatively flat. If you first draw a straight line on the pillowcase before doing this, and then place the ball near the straight line, you will see that the "line" curves due to the presence of the ball. The straight line is the path light would take in a perfectly flat spacetime, and the curved path results when light passes near a very massive object; we say that the spacetime itself is curved, just like the pillow, which causes the apparent bending of light. The space of our universe appears 3D, though, unlike the 2D surface of the pillow. However, we can illustrate the analogous effect as the warping of a 3D hyperplane. This "warped hyperplane" is a non-Euclidean spacetime. The foam of spacetime can be warped like a pillow.

Consider a distant star moving through the night sky. An occasional distant star might pass behind the sun. Of course, you won't "see" this happen in the middle of the day, but a star could be approaching (or just passing) the sun just before sunup or sundown. Such starlight would bend around the sun due to the gravitational warping of spacetime near the sun. Due to the curved paths that the starlight would take as the star passes the sun, there would be an apparent change in the star's speed through the sky as it approaches and passes the sun's angular position in the sky (of course, it's not approaching the sun itself). Precision measurements of the change in apparent speed as a star passes near the sun provide experimental confirmation for the general theory of relativity.

The closest planet to the sun, Mercury, features a noticeable – again, with precise instruments – shift in its orbit at perihelion (its point of closest approach to the sun in its elliptical orbit). As with the deflection of starlight passing near the sun, this perihelion shift can be accounted for by the curvature of spacetime in the vicinity of the sun. These are a couple of ways that the theory of general relativity has been confirmed experimentally. General relativity begins as a generalization of the special theory of relativity to the case of accelerated motion, but through the equivalence principle it is actually a theory of gravity; and such gravitational features as the warping of spacetime have been observed experimentally. Perhaps the most striking feature is the existence of black holes in our galaxy.

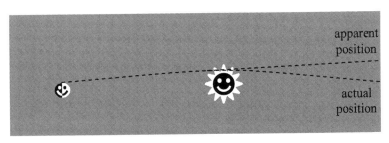

the deflection of starlight due to warped spacetime

Stars burn fuel, producing an astronomical quantity of heat (that warms planets, such as earth) and light. After billions of years, when the star has run out of fuel to burn, the star begins to collapse. This is because the process of burning thermonuclear fuel provides pressure to support against gravitational collapse. A very massive star can collapse down to a very tiny radius and form a black hole. For comparison, the entire mass of the earth would have to collapse down to the side of a pea in order to become a black hole.

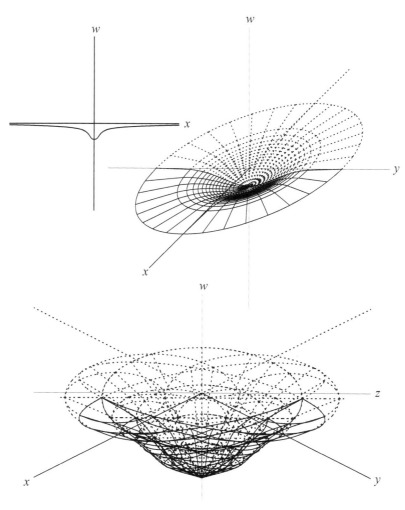

the warping of spacetime in (1+1)-D, (2+1)-D, and (3+1)-D; the (3+1)-D depression has spherical cross section, whereas the (2+1)-D depression has circular cross section

A black hole warps its local spacetime with an extreme effect. Inside the black hole, everything shrinks down to an infinitesimal point. A black hole looks somewhat similar to Gabriel's horn, except that it would be a 4D generalization of Gabriel's horn (though a black hole is generally depicted as if it were a 2D space collapsing to a point, as we can visualize this more easily in (2+1)-D spacetime – but spacetime is not (2+1)-D). Gabriel's horn is a curious geometric structure with finite volume and infinite surface area: You can pour a finite amount of paint into Gabriel's horn and fill it, but if you proceed to paint Gabriel's horn, you need an infinite amount of paint to do it. A black hole has a similar shape when drawn not in space, but in (3+1)-D spacetime.

Puzzle 10.8: Resolve the paradox with Gabriel's horn. Hint: It's *dimensional*.

Although the star collapses to a singularity – i.e. all the matter collecting at the black hole squeezes into an infinitesimal point – a distant observer sees something altogether different. Of course, you don't "see" a black hole the way you see a star, but it blocks light directly behind it and deflects light passing nearby and has additional astrophysical effects that can help to detect the black hole indirectly. A distant observer can make out an event horizon. Objects approaching the event horizon appear to get exponentially closer to the event horizon, without ever reaching it. Objects are rapidly accelerated toward the black hole, and these high-speed objects slow down and contract. So while objects approaching a black hole reach it and collapse down to a point in a finite time interval, to a distant observer it appears that they never quite get there. Even the star itself appears to collapse gradually down to a finite radius called the Schwarzschild radius, where it seems to stop. Nothing inside the event horizon can be observed from outside the black hole – a sort of "cosmic censorship." This is indeed a severe warping of spacetime in the vicinity of a black hole.

Associated with the slowing down observed for objects approaching a black hole is a gravitational redshift. When an observer in a gravitational field receives an electromagnetic signal from a source that is in a region of the gravitational field that is stronger, the received signal has a smaller frequency. This is called gravitational redshift, since red has the shortest frequency of visible light. (The antithesis is called gravitational blueshift.) There is another redshift that we observe due to the Doppler effect: A Doppler shift occurs due to a relative velocity between the source an observer. A Doppler redshift is observed when the source is getting further from the receiver.

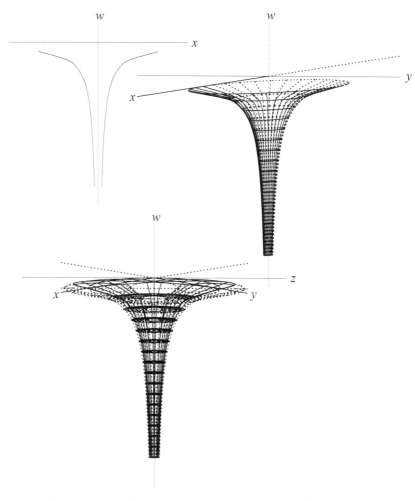

black holes depicted as a warping of (1+1)-D, (2+1)-D, and (3+1)-D spacetime; the (3+1)-D black hole has spherical cross section, while the (2+1)-D black hole has circular cross section

More than a theory of gravity, general relativity is significant cosmologically in that it can explain the origin of the universe as a Big Bang. As a theory of gravity, general relativity is fundamental to any quest for an ultimate theory of everything, for which superstring theory is presently our best candidate. Black holes show a strong connection between general relativity, quantum mechanics, and thermodynamics. A theory that unifies all four fundamental forces must reveal that gravity and the other three forces are all different manifestations of a single, underlying force. We presently understand the quantum mechanics of the other three forces, but not gravity (since it is so weak that it can only be observed macroscopically – i.e. it has not been probed at the atomic level). Black holes in the theory of general relativity (aka gravitation) provide a link between gravity and quantum mechanics – a hint, and a hope for ultimate unification. So now let us turn our attention to quantum mechanics.

10.2 Quantum Mechanics

Theories for the fundamental forces exerted between two particles inherently involve quantum mechanics. Here we survey some basic features of the quantum theory, which, like relativity, can seem very counterintuitive compared to everyday experience. We went into some conceptual detail in our introduction to relativity because the interrelation between space and time, seen in the special and general theories of relativity, are critical to realizing that, and how, the spatial extra dimensions that we have considered are part of a grander spacetime continuum. Quantum mechanics is relevant through its role in a fundamental theory of everything, for which superstring theory is our prime motivation for the physics of large extra dimensions. Still, we will keep our introduction to features of the quantum theory brief, and refer the interested reader to suggested readings at the end of the chapter.
    Quantum mechanics describes the physics of interactions between individual particles. When a block slides down an incline, the block and incline themselves consist of many times Avogadro's number of particles and the macroscopic effect that is observed – namely, the block sliding with uniform acceleration (or deceleration, depending upon the coefficient of friction), or non-uniform acceleration if air resistance is significant – ultimately depends upon the virtually countless interactions between individual particles. The block sliding down an incline is a macroscopic interaction, and we can treat it classically with virtually zero error from neglect of any quantum mechanical effects. As we shall see, microscopic interactions between individual particles may be very different from classical physics, which best describes macroscopic interactions. However, there are some quantum effects that are observed macroscopically, such as when you

shine Avogadro's number of photons on a metal and observe the photoelectric effect (to be described), so it is not quite correct to say that quantum mechanics is strictly a microscopic phenomenon. A better rule of thumb has to do with momentum, in relation to deBroglie's formula. When we get to deBroglie's formula, you'll see that quantum mechanics is particularly important for objects that do not have very large momentum. Macroscopic objects, like a block, inherently have so much mass that if they have any discernible velocity, they will have too much momentum to have any noticeable quantum effects. Now we shall survey a variety of concepts in turn.

Max Planck resolved a major discrepancy between experimental data for blackbody radiation and the classical theoretical prediction – a complex conceptual and mathematical derivation involving thermodynamics, statistical mechanics, electromagnetism, and standing waves. Planck showed the theory and experiment would agree if the energy of electromagnetic waves – i.e. light – was quantized, meaning that the energy could not equal any real number, but only integral multiples of hf, where f is the frequency of the electromagnetic wave and h is now called Planck's constant and has the value 6.626×10^{-34} J·s. This means that light of a given frequency f can have an energy hf, $2hf$, $3hf$, and so on – so it could have a frequency of $7hf$ or $196hf$, for example, but not $0.4hf$ or $14.5hf$. We say that light carries a *quantum* of energy.

Puzzle 10.9: Like energy, electric charge is also quantized. Why isn't possible to pick up an object that has a charge of $3.5e$, for example?

The classical theory also failed to explain another phenomenon called the photoelectric effect. The photoelectric effect is observed when a photon carrying sufficient energy strikes a metal surface and results in the ejection of an electron (called a photoelectron in this context) from the metal. Einstein successfully explained the photoelectric effect by interpreting Planck's solution to the blackbody radiation problem. Einstein realized that if light was made up of particles, called photons, this would naturally explain why the energy of electromagnetic waves is quantized – you can only have an integral number of photons, so to have an energy of $2.5hf$, for example, you would have to have two and one-half photons, which is absurd. The photoelectric effect occurs when an incident photon has enough energy to transfer to an atom to ionize one of its electrons. A quantum of

187

energy, hf, corresponds to a single photon. A beam of light consists of several photons, and thus carries an energy of hf times the number of photons in the beam (if you want to get picky, the source of a beam is constantly producing photons until turned off, so it makes more sense to work with the number of photons per unit time – and this rate at which energy is transferred corresponds to the power output of the beam).

Puzzle 10.10: Explain each of the following observations regarding the photoelectric effect. When the photoelectric effect was first observed, historically, these were very serious issues.

- Whether or not photoelectrons are emitted depends upon the *color* (which may not be visible) of the light.
- The maximum kinetic energy of photoelectrons is independent of the intensity of the beam – you get the same result for a very dim, weak beam as you do when you crank the intensity way up.
- However, a very dim beam with higher-frequency light results in photoelectrons with more kinetic energy, on average, than a lower-frequency beam of light that has very high intensity.
- The intensity of the beam does affect the number of photoelectrons observed (but not their kinetic energy).
- Photoelectrons are ejected from the metal almost immediately – i.e. it does not take any noticeable time for the effect to build up.

Historically, there had been a long debate over whether or not light was a wave or made of particles, and by the time Einstein came around everyone "knew" that light was wave-like. If you shine a laser beam through a couple of narrow slits, for example, it results in an interference pattern that is characteristic of wave interference – e.g. ripples in a water tank passing through narrow openings produce the same wave pattern – and not of particles (if you roll several bowling balls out a door, you won't see a pattern remotely like that of the double-slit experiment). But then Einstein solves the photoelectric effect puzzle by interpreting light as particle-like! Well, which is it?

Our modern-day understanding of the nature of light is that it has a dual nature – it has both particle-like and wave-like properties. The photoelectric effect reveals light as particle-like, while the double-slit reveals the wave nature of light. So, light, thought at the time to be purely a wave described by Maxwell's equations, turns out to both particle-like and wave-like. If you accept this, you might wonder about ordinary "particles" – like the electron. As it turns out, all "particles" have both particle-like and wave-like properties. If you direct an electron beam to pass through a pair of narrow slits, it also results in an interference pattern that is characteristic of waves.

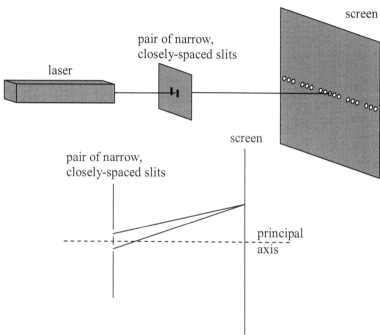

the modern-day double-slit experiment (variation of Young's experiment): the beam of light shining on the two narrow, closely-spaced slits does not pass straight through, leaving two bright spots on the screen – as might naïvely be expected – but instead results in a distinct interference pattern characteristic of waves; light traveling from one slit to a given point on the screen off the optic axis travels a different distance compared to the other slit; when this difference in optical path length equals an integral (half-integral) number of wavelengths, a bright (dark) spot appears on the screen

Combining the quantum energy of a photon, $E = hf$, with Einstein's famous equation, $E = mc^2$, results in $hf = mc^2$. Dividing both sides by c, $hf/c = mc$. The right-hand side is the magnitude of the momentum p carried by a photon – which, despite having zero-rest mass, does have relativistic mass, bending in gravitational fields, and transports energy and momentum. On the left-hand side, we can apply the equation $c = \lambda f$, which relates the speed of an electromagnetic wave to its wavelength and frequency. Doing so, we arrive at deBroglie's relation: $p = h/\lambda$.

The deBroglie equation relates the momentum of a particle to its wavelength, and applies to all "particles," not just photons. The wavelength is characteristic of the wave nature of the particle, and any particle that has momentum exhibits wave-like behavior. Since momentum and wavelength have a reciprocal relationship, a particle with less momentum (mass times velocity) is more wave-like, which explains why a bowling ball does not exhibit any discernible wave-like qualities (owing to the bowling ball's macroscopic mass, on the order of kilograms, any bowling ball moving with a noticeable speed will have so much momentum that its wavelength will be miniscule), whereas electrons have rather extreme wave-like properties (the mass of an electron is about 10^{32} times smaller, so even an electron moving nearly light-speed can exhibit very significant wave-like behavior).

When we say that particles are wave-like, we really mean that they are probability waves. When a single particle passes through a narrow slit, often it does not pass straight through, but would strike a screen off-axis. It is impossible to predict where a single particle will strike the screen, but we can compute probability distributions – i.e. we can place odds on the particle striking a section of the screen. If a beam of particles passes through a narrow slit, they fill out a pattern that matches the theoretical probability distributions quite well.

Uncertainty is inherent in quantum mechanics. Such probabilities did not appeal to Einstein, who insisted that, "God does not play dice!" Nonetheless, the theory of quantum mechanics remains a theoretical prescription for how to compute the probabilities and continues to agree with experiment, so it seems that God *might* play dice after all – or at least that particles do the equivalent. Since we can't know exactly where a single particle will strike the screen, it means that there is some uncertainty in the particle's position. Heisenberg found that the uncertainty in a particle's location is related to the uncertainty in a particle's momentum: The product of these uncertainties must be greater than or equal to $h/4\pi$. This means you can't know exactly where a particle is and exactly how fast it is moving at the same time. The more precisely you know one, the less precisely you know the other.

To get a feel for Heisenberg's uncertainty principle, imagine a macroscopic object like a ball that has a mass of a kilogram or more and a speed of a m/s or more. According to Heisenberg's uncertainty principle, the product of the uncertainties in the ball's momentum and position must exceed about 10^{-35} J·s, which is several orders of magnitude smaller than any conceivable error in measuring the ball's mass, speed, and position. Thus, we conclude that the macroscopic objects that we deal with everyday do not have any noticeable quantum effects, which is good because it means that we can reconcile how the strange effects of the quantum world differ from ordinary human experience. An electron, however, has a mass that is about 30 orders of magnitude smaller than a 1-kg ball (and can easily accelerate to high speeds, yet the uncertainty principle is significant even for low speeds), so we see that the uncertainty principle is very significant for individual particles.

The uncertainties characteristic of the quantum theory are exemplified by Schrödinger's cat. If a cat is sealed in a box in which a radioactive isotope is being released, a poison which will kill the cat, eventually, according to quantum mechanics, you can't predict with certainty whether the cat will be alive or dead if you open the lid after some time. From a theoretical perspective, we would say that the cat exists in a superposition of eigenstates – i.e. the two possibilities, alive and dead – until a measurement is made (namely, the box is opened), at which point one of the two eigenstates will be selected. You can't *know* which eigenstate will be selected, but must be content with computing the probabilities that either eigenstate will be revealed upon measurement. However, you can compare theory and experiment and obtain a very high degree of success: By repeating an experiment[15] a large number of times, the probability distributions of quantum mechanics agree remarkably well with observation. That is, you can predict that 75% of the time you will measure a particular eigenstate (e.g. the spin of the electron, which can be up or down) and 25% the other eigenstate, for a particular measurement (for many measurements, there are more than two eigenstates possible). If this is compared with 1000 observations, for this repeated experiment we expect to see

---

[15] Of course, quite humanely, we don't use lab cats in our experiments – in case you may have been wondering.

about[16] 750 observations of spin up and about 250 with spin down. If you just make a single measurement, though, we're back to square one: You can lay down the odds, but you can't be *sure*.

In first-semester physics, we teach students how to determine where a projectile will land given its initial position and initial velocity. In quantum mechanics, we can't even ask the same question, let alone work out the answer! You can't even know how fast an electron is moving and where it is to begin with, due to the uncertainty principle, so knowing exactly where it will land is totally out of the question. The best we can do is specify an electron cloud that represents possible initial positions of the electron – somewhere in the cloud – and a similar bubble of possible initial momenta, and then we can compute another cloud representing possible final positions. We inherently must work with these gray areas. The cloud represents the inherent uncertainties – which are fundamental limitations, not related to lack of better equipment – in measuring the particle's position – but we can be sure it is *somewhere* within the specified cloud.

After all, particles are partly particle-like and partly wave-like. A wave spreads out as it propagates, and is not a localized infinitesimal unit. So particles are not "pointlike" in this respect. In quantum mechanics, we use Schrödinger's equation to compute the wave function ψ of a particle, and knowledge of ψ can then be used to compute any desired probabilities and uncertainties – i.e. to specify the "cloud."

We see these features of quantum mechanics in the hydrogen atom, for example. An electron does not travel around the nucleus (which is just a proton – i.e. no neutrons – for the most abundant isotope of hydrogen) in perfectly circular orbits, or any well-defined trajectory, as we would expect by analogy with possible satellite orbits, but instead resides in one of the available subshells – an electron cloud. The electron can reside in one of various shells, which correspond to allowed energy levels, and in each shell the electron resides in one of the available subshells, which correspond to possible values of angular momentum. The electron can not have any value of energy or angular momentum – like the energy of a photon, the energy and angular momentum of a bound electron are observed to be quantized. The electron clouds are related to the wave-like nature of the electron, and represent the probabilities and uncertainties inherent in quantum mechanics. An electron can make transitions between different energy

---

[16] "About" can be quantified very well with statistical treatments, so it's not "guesswork." This is a conceptual book, so we're not going into the statistical details; "about" will have to suffice for us.

levels, either by absorbing an incident photon to jump to a higher energy level (or, if the photon has high enough frequency, the electron can leave the atom – i.e. the atom becomes ionized) or jumping down to a lower energy level by emitting a photon. We observe such transitions when we look at atomic spectra – i.e. the spectral lines that are a unique signature of an element, as seen when viewing light emitted by a gas discharge tube through a prism or grating spectrometer, for example. A discrete set of spectral lines appear, corresponding to possible transitions between energy levels. Since only certain energy levels are allowed, the number of transitions is limited. The difference between the energy levels dictates the frequency of the emitted or absorbed photon through Planck's formula ($E = hf$). Thus, atomic spectra show that energy is indeed quantized.

Some fascinating phenomena result from the probabilistic world of quantum mechanics, such as tunneling. An electron has a probability of tunneling through a barrier that would be classically forbidden for a particle; in the case of tunneling, a particle passes through a forbidden zone (a region where a particle could not exist and have a real value for energy), revealing its wave-like nature. While electrons may tunnel through barriers, you can roll a tennis ball toward a brick wall all day long for the rest of your life, and it will never appear on the other side because a moving electron has a significant deBroglie wavelength, whereas a tennis ball that is perceptibly moving has virtually no deBroglie wavelength – hence, a tennis ball (as a whole, anyway) only exhibits particle-like properties.[17]

10.3 Quantum Field Theory

Our everyday experience is with macroscopic objects that are composed of an astronomical number of particles and are incredibly slow compared to the speed of light. The effects of quantum mechanics and relativity are negligible in this limit, and the physics that describes these experiences is classical mechanics. In our quest for a fundamental theory of everything, we analyze interactions between individual particles – since, at the fundamental level, any effective interactions between conglomerate objects results from the interactions between the particles that make up the objects. So inherently a theory

---

[17] Whether or not the wall may erode after a very long period of time (or if you throw it the speed of light) is a separate matter. In that case, there will be an obvious hole in the wall. The point is that a tennis ball will never pass through the wall without leaving some sign of damage.

of fundamental forces must be a quantum theory. Additionally, since individual particles have very little mass, they are very easily accelerated. As a result, when individual particles interact, it is not uncommon for one of the particles to be moving close to the speed of light. Experimentally, we learn more about particles when they interact at very high energies, which means at least one of the particles is moving very fast. Thus, any theory of fundamental forces that we wish to compare with experimental observation must also incorporate the theory of special relativity.

Quantum field theory combines quantum mechanics with special relativity to describe interactions between individual particles in which one or both of the objects is moving close to the speed of light (either relative to the lab or their center of mass). Certainly, a theory of everything must explain both quantum and relativistic effects. Quantum field theories are rich in sophisticated mathematical treatments, though the interactions between particles can be visualized conceptually with Feynman diagrams.

The original quantum field theory was quantum electrodynamics (QED), which describes how pairs of charged particles interact electromagnetically. The quantum of electromagnetic energy is the photon. The basic QED interaction is the coupling of two charged particles to a single photon, and all purely QED processes are built from this. An electron and positron can meet and annihilate, producing two photons in the process. The reverse, called pair production, is also possible: Two high-energy photons can collide and produce an electron and positron.

The strong interaction is similarly described by quantum chromodynamics (QCD). This applies to quarks attracting one another through the strong nuclear force. Here, *chromo-* refers to color, since quarks come in one of three types of strong charge, called color – red, blue, and green. Compare to QED, where charged particles either have positive or negative electric charge. In QCD, the mediator is the gluon, which is analogous to the photon in QED. However, gluons can couple to one another – three or four gluons to a vertex – which has no analogy in QED. This is related to the fact that the strong interaction is a short-range interaction, for distances of about the size of an atomic nucleus or less, whereas the electromagnetic interaction has infinite range.

The quantum field theory for the weak interaction is quantum flavordynamics. We usually concentrate on QED, which can be extended to include the weak interaction. This is called the electroweak theory, and together with QCD forms the Standard Model for elementary particles and their interactions. Gravity is left out of the picture in the Standard Model: Its long-range form is well understood, but quantum gravity is not; the mediator of the gravitational interaction, called the graviton, has not yet been discovered. This has to do with the extreme weakness of gravity, compared to the other fundamental forces, at observable energy scales. Gravity is very strong for an astronomical amount of mass, but short-distance interactions or interactions between individual particles (the basis for quantum gravity) are extraordinarily weak. String theory is presently our best bet for unifying gravity with the other fundamental forces and understanding it at the quantum level. The recent motivation for large extra dimensions provides us with the tantalizing possibility that we will be able to understand, and detect, gravity at reasonably low energies (like that of the LHC) – i.e. low compared to the original Planck energy (see Chapter 8).

10.4 Grand Unified Theories

Two seemingly different theories become unified when experimental and theoretical developments reveal that they are two different manifestations of a single, underlying theory. We call this *unification*. Such unification has occurred in the history of physics, and serves as a hint that all four fundamental forces of nature may be seemingly unified by an ultimate theory of everything.

For example, the early Greeks made many experimental advances in the precision of solar system data, from which they were able to develop a progressively better (albeit fundamentally incorrect) model of the solar system, while also they gave some thought to a set of laws that governed motion on earth. The Greeks expected the laws of physics that governed the heavens to be different from the laws of physics that governed motion on earth. They saw these bright points of light in the night sky and mapped out their motions, and at first explained motion in the solar system in terms of circular orbits around the earth. They associated the tiny lights in the sky with perfect, shiny orbs, and the orbits as perfect circles. After all, in the heavens, every detail should be aesthetically perfect, like the circle – which they loved for its geometric simplicity and continuity. However, they could not apply the same philosophy to the obviously 'imperfect'[18] life on earth. The natural dividing line between terrestrial and celestial physics is the moon, for its visible 'imperfections.'

As they obtained better and better astronomical data, they realized that the orbits could not be perfect circles, so, in the spirit of

---

[18] Remember, it's all relative!

science, they revised their model. They were quite clever: They made circles within circles, like a Spirograph – planets orbited the earth along an *epicycle*, which itself moved along another circle called a *deferent*. It was also noted that the earth could not be at the center, so they introduced a point called an *equant* – a partner for the earth's position. Eventually, more epicycles were needed to keep up with the data, and the theory became quite complex.[19]

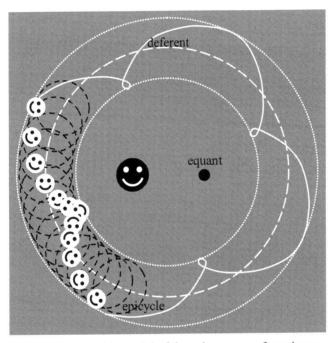

the Greek geocentric model of the solar system, featuring an earth near the center, partnered with an imaginary point called the equant: the sun and planets travel along an epicycles, a small circle that travels along a larger circle called a deferent

Meanwhile, it was observed that terrestrial objects tend to come to rest (though they did not realize that this was not their natural tendency), while objects in the heavens tend to stay in motion. But this was not a problem, since celestial and terrestrial physics were not in the

---

[19] This approximation method is analogous to the spirit of the Fourier series and renormalization – topics that we encountered previously.

first place expected to agree, and there could also be some (unknown, yet perhaps postulated) mechanism that keeps celestial objects in motion. Anyway, if a rock was sliding, it quickly came to rest, if an object was falling, it eventually struck the earth and came to rest, etc. You could pick an object up and make it move, until you yourself tired out. Though, if you don't have a firm grip, the object may fall off: If you hold your palm open with a ball of paper atop it and start running, for example, the ball of paper will fall off and wind up on the ground behind you. (Try it with a brick, though, and you may instead discover inertia, rather than air resistance.)

Thus, the Greeks (incorrectly) reasoned that the earth had to be in the center of the solar system. The heliocentric theory was considered by the Greeks, but abandoned in favor of the geocentric theory.[20] One of their arguments is that the earth would have to move across a vast expanse of space to complete its orbit in one year, for which it would have to move with such a large speed that people, animals, rocks, and all other loose objects would simply fall off – like the ball of paper that falls off your palm when you run (but what about the brick?).

These notions were challenged during the Renaissance. Copernicus proposed that the heliocentric model would offer a much simpler explanation, though there is still the issue of why we don't fall off the earth. Galileo made extensive use of the telescope (although he did not invent it), and observed mountains on the moon (which at first he thought were lakes), spots on the sun, and phases for Venus. Galileo's observations did not agree with the early Greek notions of perfectly celestial orbs. In particular, Galileo observed a full Venus, which is only possible in the heliocentric model (in the geocentric picture, the inner planets had to be constrained to lie close to the sun, unlike the outer planets, in order to explain why they are never observed too far from the horizon or too far from sunrise or sunset – of course, this is naturally explained in the heliocentric model). These scientific challenges to the Greek doctrines for astronomy and terrestrial physics paved the way for coming scientists to free themselves from the constraints of these Greek preconceptions.

---

[20] An interesting anagram is the egocentric model, in which you are the center of the universe. When others perceive you to walk east, you say, "No. I'm the center of the universe. When you think I walk east, what really happens is that I stay still while the rest of the universe moves west." Remember, it's all relative!

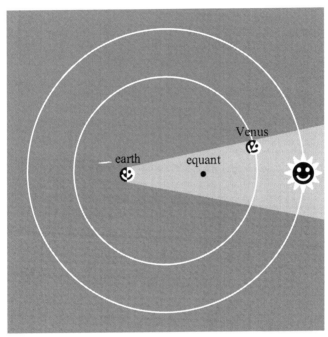

the inner planets must lie in a region near the sun in the geocentric picture in order to explain why they don't stray too far from the horizon; thus, it is not possible to observe a full Venus in the geocentric model

Puzzle 10.11. In the heliocentric picture, why can't Mercury and Venus be observed too far from the horizon or too far from sunrise or sunset? Why is there no chance of seeing them in the middle of the night in the middle of the sky? Why are there no such restrictions on the outer planets?

Sir Isaac Newton, following Galileo's work, unified terrestrial and celestial physics. Galileo demonstrated that a satellite is like a projectile that has such a high speed that it doesn't strike the ground, which showed conceptually that the physics of objects falling near earth's surface could be related to the orbits of moons and planets in the solar system. Newton modeled gravity mathematically, showing that an inverse-square law of gravity correctly explained both terrestrial motion near the surface of the earth and Kepler's laws of motion for the solar system. This is a sense of unification: Kepler's laws for solar system physics and terrestrial motion near the earth, originally thought to be two different physical laws, turned out to be different manifestations of a single underlying theory of gravity.

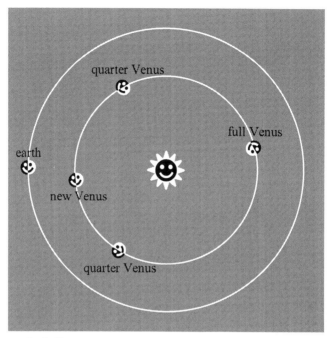

in the heliocentric picture, the fact that the inner planets can only be observed not too far from the horizon and not too far from sunrise or sunset is naturally explained without restricting the motion of the inner planets; this model is consistent with the observation of a full Venus

    Newton also explained why we don't fall off the earth even though the earth is moving 30,000 m/s in its orbit around the sun: Objects have inertia. The earth, and all of the objects on the earth, are all moving 30,000 m/s, and have a natural tendency to stay in motion. The Greeks incorrectly believed that objects had a natural tendency to be at rest. Terrestrial objects come to rest due to frictional forces: A moving object has a natural tendency to stay in motion, but slows down if a net force acts to decelerate it. This is easy to see from the perspective that there is a single explanation for the motion of terrestrial and celestial objects, since we can now clearly see that celestial objects have a natural tendency to stay in motion (there is evidently no 'mechanism' to keep them moving) – as many have apparently been in motion for millions and millions of years.[21]

There was a similar unification of electricity and magnetism into a single, underlying electromagnetic interaction, as we discussed in Chapter 7. The Standard Model of elementary particles and their interactions shows a very formal, mathematical unification of two of the fundamental forces – the electromagnetic and weak interactions (Chapter 9). We now know that the electromagnetic interaction and weak interaction are two seemingly very different manifestations of a single, underlying electroweak interaction, and that the difference has to do with the breaking of a symmetry in the underlying theory (associated with the Higgs mechanism, but while unification and its effects have been observed, the Higgs, as of yet, has not).

These historical breakthroughs in the development of our understanding of physics motivate our quest for an ultimate unification of all four forces – i.e. we anticipate a single fundamental force, not four *fundamental* forces. Quantum field theories that aim to provide further unification, compared to our present understanding, are called Grand Unified Theories. While the math can be very sophisticated, we can understand some of the features conceptually, such as some elements of the group theory.

Group theory is a branch of mathematics that describes, in very general terms, transformations of elements that form a group. It is very practical for groups with symmetry and invariance properties, and provides a mathematically formal means of quantifying conceptual aspects and properties of groups. For example, crystals have structures that are symmetric for various rotations and/or inversions, and a particular crystal structure can be described by a specific

---

[21] Many fault the Greeks for developing an incorrect notion for inertia, but they did correctly observe that objects near earth tend to lose speed. The same critics often fault the Greeks for thinking that heavy objects fall faster than light objects – in particular, that the Greeks were aesthetic philosophers and not scientific experimentalists. After all, if they simply picked up a light and heavy rock, they would see that they struck the ground at nearly the same time. Yet, it is possible that the Greeks did this experiment, but with a feather and a rock. They may have based terrestrial motion on practical observations of the effects of friction and air resistance, whereas the modern view of physics is a hypothetical framework of a perfect vacuum. Of course, this viewpoint is necessary to correctly develop an understanding of inertia, but our regular experience with air resistance and friction often causes applications of inertia to be very counterintuitive for students.

transformation group. The distance between neighboring atoms would remain invariant under rotations and reflections. Such symmetries and invariance can be exploited mathematically in the context of group theory to calculate useful properties of the crystal.

It turns out that elementary particles and their interactions can be described by groups that can also describe the rotation of objects in N-dimensional space. We saw in Chapter 9 that the $W^{\pm}$ bosons couple only to the left-chiral fermions. These interactions can be described by the group $SU(2)_L$, and we tack on the 'L' to designate that it does not apply to right-chiral fermions. For example, the u and d quarks form a left-chiral doublet, as the left-chiral u and d can couple directly to one another via a $W^{\pm}$; the left-chiral e and v_e form a similar doublet. However, their right-chiral counterparts can not couple to one another directly. The right-chiral fields form the group $U(1)_R$. The electromagnetic interaction couples pairs of identical particles (or particle-antiparticle pairs) through the interaction $U(1)_{EM}$. In the electroweak theory, where the electromagnetic and weak interactions are unified, the electroweak interaction is described by $SU(2)_L \times U_Y(1)$. The Y designates hypercharge, which is a combination of the particle's electric charge and weak isospin (the "weak charge" – each force has its associated charge; cf. strong charge, which we call color). The strong interaction is described by $SU(3)_c$, where the c represents color. The Standard Model for elementary particles and their interactions is represented by $SU(3)_c \times SU(2)_L \times U_Y(1)$. This is not a Grand Unified Theory, but represents the current state of the theory that has been verified by experiment.

The unification that developed on the road to the Standard Model suggests that there may be a grander unification. Grand Unified Theories work with larger groups, like $SU(5)$ and $SO(10)$,[22] which are expected to break down to the Standard Model, $SU(3)_c \times SU(2)_L \times U_Y(1)$, plus additional group(s), via some symmetry-breaking (like the mechanism that breaks symmetry in the Standard Model at electroweak energies, causing a single interaction to manifest itself as either an electromagnetic or weak interaction

---

[22] The S stands for special, meaning that the determinant of matrices representing the elements of the group are plus or minus one, while U and O stand for unitary and orthogonal, which are particular types of operators in the mathematical field of linear algebra.

depending upon the circumstances). Such Grand Unified Theories generally predict the existence of particles that have yet to be detected experimentally, since larger groups hold extra content. For this to be viable, these extra particles must be heavy enough to explain why they have not yet been produced at high-energy colliders. By the same token, we expect to discover heavy particles at the LHC if indeed there is a grander unification.

Let's look at one possible higher symmetry now, and then we'll look at another in a later section (namely, supersymmetry). Grand Unified Theories will hopefully answer several fundamental questions, for which there currently is no confirmable explanation. For example, why are there so many particles with different masses, why do we see much matter in the universe, but comparatively few antiparticles, and why do right-chiral fields couple differently than left-chiral fields? Regarding this last question, it is alluring to develop a Grand Unified Theory that contains not only $SU(2)_L$ and $U(1)_R$, but also $SU(2)_R$ and $U(1)_L$. This puts the left- and right-chiral fields on an equal footing in the Grand Unified Theory, and is what would be observed at a high enough energy scale. This is termed the *unification scale* – i.e. at a high enough energy, the fundamental forces unify into a single force, but diverge into different manifestations at low energies. The left-right symmetry is broken below this energy scale, and this symmetry-breaking is used to explain our low-energy observations. This is the essence of how these symmetry arguments go.

10.5 String Theory

When we wonder what matter is made of, we generally mean to resolve whether it is composed of discrete pointlike particles or is continuous – i.e. infinitely divisible. The Greeks wrestled with this question philosophically, favoring the notion of the *atom* – not quite what we call an atom, but some basic unit of matter that is not divisible. More recently in history, with the beginnings of chemistry, we discovered the elements. These were so named because they were thought to be the most fundamental particles of matter – what the Greeks called atoms. The elements turned out to be different types of atoms, but atoms are not elementary – they are composed of protons, neutrons, and electrons. The protons and neutrons turned out to be composed of quarks, but it appears that quarks, electrons, and other particles tabulated in Chapter 9 are finally elementary particles. But are elementary particles pointlike?

Well, we already answered this question in the section on quantum mechanics: All particles exhibit both particle-like and wave-like properties. So they aren't pure point-particles in this regard. It makes you wonder exactly what a "particle" is. We know that particles have a property called mass, which is a measure of inertia. Some particles, like the photon and gluon, carry zero rest-mass, but always travel the speed of light and so are never at rest, yet still have mass in the sense that they have relativistic inertia. The elementary particles come in a variety of masses that do not appear to fit into a simple pattern (though there is an abundance of literature geared toward explaining the mass hierarchy). Each particle may also carry electric charge, strong charge, and isospin. Mass plays two roles, in that it serves as inertia in addition to "gravitational charge." Particles also have spin angular momentum, which can come in half-integral units (fermions) or integral units (bosons), including the case of zero spin. Also, there are a large number of particles – three generations of up-type quarks, down-type quarks, charged leptons, neutrinos, and their mediators, with different behavior between left- and right-chiral fermions. There sure are a lot of "elementary" particles, which come in a variety of combinations of parameters corresponding to different aspects of particle behavior.

In our quest for an ultimate theory of everything to unite the four "fundamental" forces, we hope that the theory will also explain the number of particles and their properties. Perhaps all of these "elementary" particles will turn out to be different manifestations of a smaller set of elementary objects.

Let's return to something we mentioned a moment ago: Particles have mass and come in a wide range of rest-masses. Recall that mass is equivalent to energy through Einstein's famous equation. Why does a particle at rest have energy? We know that it does. For example, an electron and positron at rest (or nearly so) in close proximity can interact and produce a pair of photons, which have zero rest-mass. These particles of matter are converted to electromagnetic radiation, which shows that rest-mass is just a form of energy.

Particles at rest carry energy, and particles exhibit wave-like properties. Could it be that particles are vibrations of some sort? Say, vibrating strings, or vibrating drums (a 2D rather than 1D medium), or vibrations of the spacetime itself – a sort of spacetime "foam" defining the geometry of the universe. Perhaps particles are resonances – i.e. standing waves. We know that if we vibrate a string that we can produce standing waves on the string if the frequency matches the length of the string just right, and that these vibrating strings carry energy. Perhaps particles are resonating strings.

An ordinary string of a given length can produce more than one type of standing wave. There is a fundamental resonance, a first overtone, a second overtone, and so on. Each standing wave corresponds to a different energy level. So, perhaps the variety of elementary particles observed are different types of vibrating strings, and perhaps their properties are also associated with different ways that a string can vibrate. The analogy with strings opens up some possibilities. A string can have its ends free or fixed, and the standing waves look different depending upon these boundary conditions. Analogously, in string theory, there are open and closed strings.

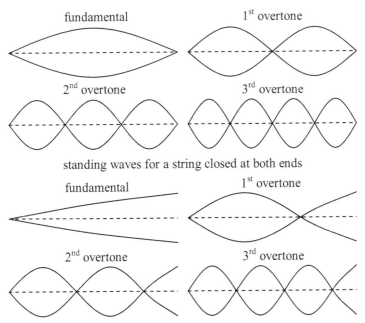

standing waves for a string closed at both ends

standing waves for a string with one end closed and one end open

Conceptually, we can see how the idea that particles may actually be vibrating strings may come about, and explore some of the theoretical implications that it has, but string theory is a very sophisticated mathematical theory. The details come from solving these mathematical equations, and the possibilities are limited by mathematical constraints. While string is a theory of everything, it is different from the traditional Grand Unified Theories in that it involves a completely different theoretical approach. Grand Unified Theories aim to extend the group content of Standard Model, while string theory starts at the fundamental level with the concept of a string (yet, string theory will ultimately involve a larger group content to hold all of its particles, of which the Standard Model will be obtained a subgroup, likely after some type of symmetry breaking).

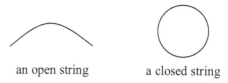

an open string   a closed string

String theory is currently our best candidate for a theory of everything. It can incorporate gravity and the other three "fundamental" forces together. The gravitational part includes general relativity with quantum gravity, in terms of world sheets rather than world lines. A point-particle would sweep out a world line as it evolves through spacetime; a string sweeps out a world sheet. We draw Feynman diagrams for the interactions of point-particles as stick-like figures. However, in string theory, the elementary particles are strings, not points, and so the Feynman diagrams are sheet-like rather than stick-like.

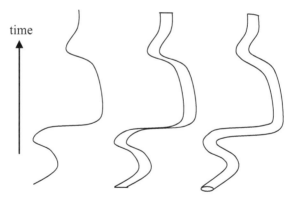

a world line for a point-particle and world sheets for open and closed strings

At the Planck scale, particles reveal themselves as strings, and there is a single fundamental force of nature. At energies below the Planck scale – and we observe this exclusively – the theory is broken, associated with some symmetry, and the single force diverges into four different possible manifestations. For a brief moment after the Big

Bang, string theory would have governed in its most fundamental form at the high energy of the Planck scale, and all of the dimensions of spacetime would have been on an equal footing (as what we would normally term the *extra* dimensions had not yet compactified).

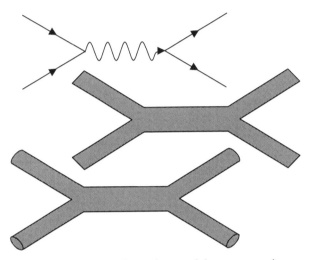

Feynman diagrams for point particles, open strings, and closed strings

The mathematics of string theory blossoms from the simple idea of a string, in analogy with physical strings on which we can produce standing waves, but applied in the context of the mathematics of quantum field theory. From this simple starting point, string theory developed into a theory that could describe the four "fundamental" forces, unifying gravity with the other three forces. String theory also shows the potential to answer some fundamental questions, which are not addressed by the Standard Model, such as why there are three generations of fermions and why the masses do not seem to follow a very simple pattern. The theory can also accommodate a Grand Unified Theory group structure, which shows the potential to lead to the Standard Model at low energies – i.e. it appears to be phenomenologically viable to obtain, starting from string theory, all of the known particles (in addition to some exotic particles that apparently have very high mass and therefore have not yet been detected). Any theory of everything that can't reproduce the known particles can immediately be ruled out, so it's important that string theory show strong promise for reproducing the Standard Model at low energies.

String theory was not well-accepted in its early stages. For one, the mathematical theory was only viable for 26-dimensional spacetime, which was perceived as a serious problem at first, since the universe appears to have three dimensions of space and one of time from our macroscopic perspective. However, string theory has made much progress. There is motivation for compactification, which resolves the issue of where the extra dimensions are. The theory has become more aesthetically appealing through theoretical developments, and some early objections have been overcome. Also, there is motivation for large extra dimensions, which addresses one of the strong criticisms that string theory was not physically testable – now, potentially, string theory can be discovered in the next generation of high-energy colliders.

The original string theory suffered from a mathematical renormalization problem associated with the hierarchy described in Chapter 8. This was solved through the introduction of a novel symmetry called supersymmetry. The inclusion of supersymmetry in string theory is called superstring theory. We describe this in the next section.

10.6 Supersymmetry and Superstrings

In Chapter 9, we learned that all particles can either be classified as fermions or bosons. Fermions are particles with half-integral spin, whereas bosons have integral spin. The quarks, charged leptons, and neutrinos are examples of fermions, and the mediators (photon, gluons, graviton, $W^{\pm}$, and Z) are bosons. These are not the only fermions and bosons, of course. For example, mesons, consisting of quark-antiquark pairs are bosons, while baryons, consisting of three quarks or three antiquarks, such as the proton and neutron, are fermions.

The notion of supersymmetry is this: For every elementary fermionic particle, there exists a bosonic counterpart, and for every elementary boson, there exists a corresponding fermion. For example, the selectron is a particle that has the same properties as an electron, except for having a much larger mass (the supersymmetric particles must be quite massive to have escaped detection from previous and current collider data) and for having integral, rather than half-integral, spin. There is a set of charged sleptons (selectron, smuon, and stau) that are perfectly analogous to the charged leptons (electron, muon, and tau) – again, except for being bosonic instead of fermionic and having more mass. Similarly, there is a corresponding set of sneutrinos and squarks. The gauge bosons also have fermionic counterparts: the photino, gluinos (gluons come in a variety of color combinations, so there are multiple kinds of gluons and gluinos), gravitino, Wino, and Zino. Sfermions, which are bosonic, tend to begin with an *s*, while the fermionic counterparts to the bosons end with *-ino*.

A natural aesthetic question to ask is why the supersymmetric particles are very massive – i.e. why don't they have the same mass as their Standard Model counterparts? (Phenomenologically, if they exist, they must be massive or we would have observed them already, but that explains why we need them to be very massive, not what *causes* them to be very massive.) This may be associated with supersymmetry breaking at sufficiently low energies. A fundamental theory is most aesthetic at a very high fundamental energy scale, and some of the underlying symmetries that exist at the fundamental scale are broken at the low energy scales presently available to us. Higher and higher energy colliders help us probe higher energy scales.

In principle, supersymmetry is a theoretical possibility whether particles are indeed pointlike particles or superstrings, but in the case of superstring theory supersymmetry is actually theoretically *motivated* as a solution to a renormalization problem associated with the hierarchy. This is a very appealing feature of superstring theory – i.e. the fact that it is renormalizable.

Returning to particles for a moment, the quantum field theory for the Standard Model is renormalizable. There are some infinities that appear in calculations corresponding to loop-level Feynman diagrams, which naturally were a major concern for the viability of quantum field theory until it was discovered that these infinities could be treated through some redefinitions of parameters of the theory – this process is called renormalization. However, these redefinitions at the one-loop level introduce new infinities at the two-loop level, so an order-by-order process is needed to completely renormalize the theory.

For the most part, string theory is renormalizable and does not suffer from problems associated with these potential infinities. We can appreciate this from a naïve perspective as follows: In quantum field theory, the infinities ultimately arise from the fact that a point-particle is a singularity; in the Feynman diagram for a point-like particle, the interactions occur at vertices which are infinitesimal points. In string theory, Feynman diagrams are sheet-like instead of stick-like, and the vertices are smooth geometric intersections rather than infinitesimal points. However, the proof is in the mathematics. Mathematical anomalies at the loop level have posed problems for alternative theoretical efforts to develop a theory of everything (like simply extending gravity to higher dimensions and trying to unify it with the other forces in the case of point-particles). However, there is a mathematical anomaly in string theory regarding the hierarchy problem described in Chapter 8, which posed a significant problem at first, but it turns out that it can be solved through the introduction of superstrings into the theory.

In string theory, there are both fermionic and bosonic strings, rather than "particles" in the usual sense. If you count, you will see that there are apparently more fermions than bosons – unless you add supersymmetry. String theory has to contain particles that become the observed particles at the low energies of the Standard Model – if it is to be at all viable – but it may also contain extra particles, as long as they are heavy enough to explain why they haven't been detected yet. At some point in the future, though, colliders will probe high enough energies that we will either discover supersymmetric particles or rule it out as a possibility. Presently, though, there is room in the theory for a host of theoretical particles with heavy masses, since we haven't yet probed higher energies.

The importance of supersymmetry is that by matching up a fermionic counterpart for every ordinary bosonic string and a bosonic counterpart for each ordinary fermionic string – hence, making superstrings – the mathematical anomalies associated with the hierarchy problem vanish. This makes a viable superstring theory out of string theory. It turns out that by doing so, the spacetime needs to be 10-dimensional instead of 26-dimensional. We "lose" 16 dimensions by adding supersymmetry.

Of course, the 6 extra dimensions must be hidden in some way, such as through compactification. The Calabi-Yau manifold, in particular, is a popular compactification scheme, which is a rather complicated topology. String and superstring theory show that theoretical constructions are related to the number of dimensions of spacetime. There is still an aesthetic hierarchy problem – instead of mathematical anomalies, we get very delicate cancellations – the fact that there are cancellations, instead of infinities, makes the theory mathematically viable, but there is still the conceptual hierarchy problem that we explored in Chapter 8. We can explain the remaining hierarchy problem with large extra dimensions – i.e. much larger than the Planck scale. It is fascinating how the number and size of the extra dimensions plays such a strong mathematical and conceptual role in the theories. We will see one more example of this in the final section.

10.7 M Theory

As string and superstring theories developed, four classes of strings were found to produce viable theories of nature. Viable means that the theory must be free of anomalies, among other things. There are other types of anomalies to consider besides the main one, mentioned earlier, which can be avoided by introducing supersymmetry. Only a string theory that can include supersymmetry, and which does not have other anomalies, is potentially viable. The four plausible classes are Type I, Type IIA, Type IIB, and heterotic strings. The heterotic strings may also be divided into two classes – two possible gauge groups to accommodate the "particles" in the theory – so in this regard, you can count five, instead of four, types of strings.

With superstrings in mind, we can describe the differences between these models in terms of two fundamental 10D[23] spinor fields. In the usual (3+1)-D spacetime, a spinor is a particle, such as the electron, which can be left- or right-chiral. The two 10D spinors in this context are not electrons or ordinary particles, though, but retain this sense of chirality associated with intrinsic spin angular momentum. Another difference between these models is the nature of the strings, which can be open or closed.

Specifically, Type I strings include open strings where the two 10D spinors have the same chirality, but doing so requires also including closed strings in order to make the theory viable – so the Type I model has both open and closed strings. Type II string theory includes only closed strings, and there are two viable ways to do this: Type IIA, in which the two 10D spinors have opposite chirality, or Type IIB, where they have the same chirality. The heterotic string has a curious hybrid nature: It splits the closed string into left and right modes, in which the right-moving modes live in a supersymmetric 10D spacetime, while the left-moving modes are bosonic only and reside in a 26D spacetime, where the extra 16 dimensions are compactified from the other 10.

These are four (or five, if you count two types of heterotic strings) fundamentally different types of string theory. It is intuitive to expect one of two things: (1) One type of string theory describes the physical universe or (2) none of these turns out to agree with experiment, if at some point we gather experimental data that can test these models. One wouldn't expect all four models to appear in the ultimate theory of everything. However, it was shown that the different types of string theory can be related by duality: A description from one model can be related to its dual model by a duality transformation. Incorporating this sense of duality into the picture and working with all of these types of string theory results in what is termed *M theory*. M theory is an effectively 11D, rather than 10D, spacetime, where the

---

[23] Meaning dimensions of spacetime.

extra dimension (compared to 10D superstrings) derives from this sense of duality that relates the various models. M theory is in a sense more fundamental than any of individual types of superstrings alone, in that it unifies the different types of superstrings – yet another, in a sense even grander, form of unification.

The reader who is hungry for more details on string theory is referred to the suggested readings [A7].

Further Reading:

There are numerous accessible introductions to the various subjects discussed in Chapter 10. The suggestions which follow are intended as a focused list to help readers find additional references on this subject, and is certainly not meant to serve as a comprehensive list. Hopefully, this will be somewhat useful by not coming across as overwhelming.

Consult [A3] for introductions to special relativity, especially the book by Einstein himself. Rudy Rucker's classic book on extra dimensions also has geometric chapters on spacetime [A4]. As the mathematics of special relativity is rather accessible, it is worth trying some textbooks on the subject, or at least the chapter devoted to this in a general physics textbook. General relativity, on the other hand, is loaded with elegant mathematics, so [A5] consists of more conceptual accounts. There are numerous works in the area of quantum mechanics, which we explored briefly, such as [A6]. Finally, for a short list of suggested readings in string theory and supersymmetry, see [A7].

Volume 2: The Physics of the Fourth and Higher Dimensions...

Volume 1 Contents

Chapter 0: The Known Dimensions

Chapter 1: Common Objections

Chapter 2: Visualizing the Fourth Dimension

Chapter 3: Higher-Dimensional Polytopes

Chapter 4: Curved Hypersurfaces

Chapter 5: A Hypothetical Hyperuniverse

References and Further Reading

These references are divided into two categories: References beginning with an A, such as [A3], are reasonably accessible to a general interest audience; those beginning with a T, as in [T2], are highly technical papers. The technical references have been kept to a minimum and are included primarily to pay tribute to a few researchers who have motivated modern-day experimental searches for large extra dimensions.

Accessible References
A1. "Large extra dimensions: A new arena for particle physics," N. Arkani-Hamed, S. Dimopoulos, and G.R. Dvali, *Phys. Today* 55N2, 35, 2002.
A2. *The Particle Garden: Our Universe as Understood by Particle Physicists*, Gordon Kane, Helix, 1996; *Deep Down Things: The Breathtaking Beauty of Particle Physics*, Bruce A. Schumm, Johns Hopkins, 2004.
A3. *Space and Time in Special Relativity*, N. David Mermin, Waveland, 1989; *Special Theory of Relativity*, David Bohm, Routledge Classics, 2006; *Relativity: The Special and General Theory*, Albert Einstein and Nigel Calder, Penguin, 2006.
A4. *The Fourth Dimension: A Guided Tour of the Higher Universes*, Rudy Rucker, Houghton Mifflin, 1984.
A5. *General Relativity from A to B*, Robert Geroch, University of Chicago Press, 1981; *Was Einstein right? Putting General Relativity to the Test*, 2nd edition, Clifford M. Will, Basic Books, 1993; *A Brief History of Time*, Stephen Hawking, Bantam, 1998.
A6. *The Quantum World: Quantum Physics for Everyone*, Kenneth W. Ford and Diane Goldstein, Harvard, 2005; *Quantum Enigma: Physics Encounters Consciousness*, Bruce Rosenblum and Fred Kuttner, Oxford, 2008; *Quantum: A Guide for the Perplexed*, Jim Al-Kahili, WN, 2004.
A7. *The Elegant Universe: Superstrings, Hidden Dimensions, and the Quest for the Ultimate Theory*, Brian Greene, W.W. Norton, 2003; *The Fabric of the Cosmos: Space, Time, and the Texture of Reality*, Brian Greene, Vintage, 2005; *The Trouble with Physics: The Rise of String Theory, the Fall of a Science, and What Comes Next*, Lee Smolin, Mariner, 2007; *Warped Passages: Unraveling the Mysteries of the Universe's Hidden Dimensions*, Lisa Randall, Harper Perennial, 2006; *Hyperspace: A Scientific Odyssey through Parallel Universes, Time Warps, and the 10th Dimension*, Michio Kaku, Anchor, 2005;

Superstrings: A Theory of Everything? P.C.W. Davies and Julian Brown, Cambridge, 1992; *Supersymmetry: Unveiling the Ultimate Laws of Nature*, Gordon Kane, Basic Books, 2001.

Technical Papers

T1. "Generalizing cross products and Maxwell's Equations to universal extra dimensions," A.W. McDavid and C.D. McMullen, hep/ph-0609260, 2006.

T2. "The hierarchy problem and new dimensions at a millimeter," N. Arkani-Hamed, S. Dimopoulos, and G.R. Dvali, *Phys. Lett.* B429, 263, 1998; "Phenomenology, astrophysics and cosmology of theories with submillimeter dimensions and TeV scale quantum gravity," N. Arkani-Hamed, S. Dimopoulos, and G. Dvali, *Phys. Rev.* D59, 086004, 1999.

T3. "A possible new dimension at a few TeV," I. Antoniadis, *Phys. Lett.* B246, 377, 1990; "New dimensions at a millimeter to a Fermi and superstrings at a TeV," I. Antoniadis, N. Arkani-Hamed, S. Dimopoulos, and G. Dvali, *Phys. Lett.* B436, 257, 1998.

T4. "Bounds on universal extra dimensions," T. Appelquist, H.-C. Cheng, and B.A. Dobrescu, *Phys. Rev.* D64, 035002, 2001; "Bosonic supersymmetry? Getting fooled at the LHC," H.-C. Cheng, K.T. Matchev, and M. Schmaltz, *Phys. Rev.* D66, 056006, 2002.

T5. "Tests of the gravitational inverse-square law below the dark-energy length scale," D.J. Kapner *et al.*, *Phys. Rev. Lett.* 98, 021101, 2007; "Sub-millimeter tests of the gravitational inverse-square law," C.D. Hoyle *et al.*, *Phys. Rev.* D70, 042004, 2004; "Tests of the gravitational inverse-square law," E.G. Adelberger, B.R. Heckel, and A.E. Nelson, *Ann. Rev. Nucl. Part. Sci.* 53, 77, 2003; "Submillimeter tests of the gravitational inverse square law: a search for 'large' extra dimensions," C.D. Hoyle *et al.*, *Phys. Rev. Lett.* 86, 1418, 2001.

T6. "The review of particle physics," C. Amsler *et al.*, *Phys. Lett.* B667, 1, 2008. Available at pdg.lbl.gov/.

Puzzle Solutions

Answers/solutions to selected problems follow.

Puzzle 6.3: 1 m, downward. If you were to add the "missing" vector, the sum of the vectors would be zero, so the four vectors shown must be equal and opposite to what is missing.

Puzzle 6.4: 13.

Puzzle 6.5: 10.

Puzzle 6.6: Negative z-axis, in the xy plane, in the yz plane.

Puzzle 6.8: 4 m, 16 m^4, 64 m^3.

Puzzle 6.9: None!

Puzzle 6.10: The right end: The torques, but not the weights, are balanced.

Puzzle 6.11: 1, 0, 0, 0, $\hat{i}$, $-\hat{i}$.

Puzzle 7.1: None!

Puzzle 7.2: None!

Puzzle 7.3: North of the bull's-eye.

Puzzle 7.4: 9.8 m/s^2, downward. Gravitational acceleration is uniform – i.e. constant in both magnitude and direction. Acceleration is a measure of how velocity is changing, so if you thought it was zero at the top of the trajectory, let me ask you this: If the rock is momentarily at rest at the top and if it also had no acceleration, meaning that it's not changing speed, would it ever come back down?

Gravitational acceleration is constant and acts downward for the entire trip. On the way up, the rock loses speed, at the top the rock has just run out of speed and is beginning to gain speed in the opposite direction, and on the way down the rock gains speed in the downward direction. The magnitude and direction of the rock's acceleration are constant throughout the trip.

Puzzle 7.5: Neither!

Puzzle 7.7: 150 lbs. (It's not accelerating.)

Puzzle 7.9: Only if they have the same potential, in which case they are really part of the same equipotential surface. Two *different* equipotential surfaces certainly can't intersect.

Puzzle 7.10: E.

Puzzle 7.11: Into the page at X, out of the page at Y and Z.

Puzzle 7.12: Into the page at X, out of the page at Y. Note that the electron is negatively charged.

Puzzle 7.13: At the very front of the loop, the current runs to the right.

Puzzle 7.14: Straight upward, bisecting the diagram.

Puzzle 7.15: As the nucleus contains more and more positive charge, there is more electromagnetic repulsion. Also, going downward in the periodic table, the nucleus grows in size. The extra neutrons are needed to provide more attraction via the strong nuclear force without affecting the electromagnetic repulsion between protons in order to make such nuclei reasonably stable.

Puzzle 7.16: Electromagnetic.

Puzzle 8.1: The other points are located near the South Pole. Not too far from the South Pole, there is a circle of latitude that has a circumference of exactly 1 mile. So 1 mile above this circle of latitude, you can walk 1 mile south, 1 mile east, and 1 mile north and return to your starting point. A bit closer to the South Pole, there is a circle of latitude with a circumference of half a mile, in which you could walk around the circle twice. And so on.

Puzzle 9.2: Applying the simple rule given in Chapter 9, $u_1^{\bullet}\overline{u}_4^{\bullet}g_5^*$, $g_2^*g_5^*g_3^*$, and $g_2^*g_6^*g_5^*g_1^*$. This is because $1 + 4 = 5$, $2 - 5 = -3$, and $2 - 6 = -5 + 1$.

Puzzle 9.3: $t_1^{\bullet} \to b_1^{\bullet} + W^+$, $c_1^{\bullet} \to s_1^{\bullet} + W^+$, and $t_1^{\bullet} \to t + \gamma_1^*$ conserve Kaluza-Klein number. However, these will be kinematically suppressed unless there are loop-level mass splittings that make the initial particle heavier than the sum of the final state particles.

Puzzle 10.1: 50 mph, north.

Puzzle 10.2: Hint: Downwind and upwind cancel out, but crosswind adds distance each way.

Puzzle 10.3: Yes. Not without superluminal velocities.

Puzzle 10.4: Travel nearly light-speed relative to earth for this period of time. Not without superluminal velocities.

Puzzle 10.5: This is the famous twin paradox. In order to make a face-to-face comparison, one of the chimpanzees must accelerate. This chimpanzee will not be an inertial observer during the acceleration. To account for the acceleration, we should look to general relativity.

Puzzle 10.6: This is the famous garage paradox, where length contraction makes this possible. Observe that a great feature of relativistic thought experiments is that you aren't constrained by a budget, you have all the time in the world to develop the materials you need, and the equipment is always an experimentalist's dream come true.

Author's Qualifications

Chris McMullen has published the following journal articles on the collider phenomenology of string-inspired, large extra dimensions:

1. "A Mechanism for Kaluza-Klein number violation in universal extra dimensions," C.D. McMullen and S. Nandi, *J. Phys. G: Nucl. Part. Phys.* 35, 095002, 2008.
2. "Collider implications of a non-universal Higgs," C.D. McMullen and S. Nandi, *Phys. Rev.* D75, 095001, 2007.
3. "Collider implications of multiple non-universal extra dimensions," R. Ghavri, C.D. McMullen, and S. Nandi, *Phys. Rev.* D74, 015012, 2006.
4. "Collider implications of models with extra dimensions," C. Macesanu, C.D. McMullen, and S. Nandi, *Amsterdam, 2002*, ICHEP, 764, 2002.
5. "New signal for universal extra dimensions," C. Macesanu, C.D. McMullen, and S. Nandi, *Phys. Lett.* B546, 253, 2002.
6. "Collider implications of universal extra dimensions," C. Macesanu, C.D. McMullen, and S. Nandi, *Phys. Rev.* D66, 015009, 2002.
7. "Collider implications of Kaluza-Klein excitations of the gluons," D.A. Dicus, C.D. McMullen, and S. Nandi, *Phys. Rev.* D65, 076007, 2002.

The author has taught three intense 30-hour winter courses on the subjects of the fourth dimension and string-inspired extra dimensions at the Louisiana School for Math, Science, and the Arts.

Chris McMullen teaches physics at the Louisiana School for Math, Science, and the Arts – a unique specialized school for high-aptitude, high-achieving students from across the state, where two-thirds of the faculty hold a Ph.D. in their subject area. Students who opt to continue their studies in-state have many opportunities to transfer up to two years of their collegiate course work.

The author is also an adjunct physics instructor for Northwestern State University of Louisiana and began his teaching career at Penn State Altoona.

Made in the USA
Lexington, KY
29 November 2011